Springer Tracts in Modern Physics

Volume 278

Springer Tracts in Modern Physics provides comprehensive and critical reviews of topics of current interest in physics. The following fields are emphasized:

- Elementary Particle Physics
- Condensed Matter Physics
- Light Matter Interaction
- Atomic and Molecular Physics
- Complex Systems
- Fundamental Astrophysics

Suitable reviews of other fields can also be accepted. The Editors encourage prospective authors to correspond with them in advance of submitting a manuscript. For reviews of topics belonging to the above mentioned fields, they should address the responsible Editor as listed in "Contact the Editors".

More information about this series at http://www.springer.com/series/426

Jung Hoon Han

Skyrmions in Condensed Matter

Jung Hoon Han
Department of Physics
Sungkyunkwan University
Suwon
Korea (Republic of)

ISSN 0081-3869 ISSN 1615-0430 (electronic)
Springer Tracts in Modern Physics
ISBN 978-3-319-88741-8 ISBN 978-3-319-69246-3 (eBook)
https://doi.org/10.1007/978-3-319-69246-3

Softcover reprint of the hardcover 1st edition 2017

Printed on acid-free paper

This Springer imprint is published by Springer Nature
The registered company is Springer International Publishing AG
The registered company address is: Gewerbestrasse 11, 6330 Cham, Switzerland

For Jemme

Preface

It has been two decades since David Thouless's book, "*Topological Quantum Numbers in Nonrelativistic Physics*" appeared in print. I was among the handful of students lucky enough to have sat in the classroom as he presented the materials of his evolving book as lectures to graduate students. The memory of the experience has planted in me an idea to write a book of my own someday. Luckily enough, this would be it.

The bubble memory devices of the 1960s and 1970s must have presented great excitement to the scientists of the era, but like relics buried beneath the ground and forgotten, they remained little known to posterity until their re-incarnation in MnSi and other chiral hosts in 2009. Together with the revival of the science came a new name: the "skyrmion." In the bubble memory days, topology was hardly on anyone's lips; today, it is the rote jargon. Indeed, ignorance of the Berry phase no longer serves as a good excuse because it is now the stuff of graduate, and even undergraduate, curricula. A book that can help material science-oriented students to navigate the (old) sea of skyrmions in the (modern) ship of topology seems to be in order.

While the original intent was to explain skyrmions, the book appears to have evolved into a textbook that tries to instill the modern tool-sets of condensed matter theory, using the science of skyrmions as a platform and an excuse. In truth, many of the theories I discuss in the book have been available for decades, but given the popularity of skyrmions in recent years, those theories can benefit from a certain amount of refurbishing in their manner of presentation. One can thus use this book as a reference book on skyrmions in chiral magnets, or on the "theoretical essentials" that happen to have been invested in the study of skyrmions. Either way, the book would see its purpose fulfilled.

Quite a few heavyweight expositions on the field-theoretical approach to condensed matter problems are in circulation. The level of field theory used in this book is no match to those volumes. This is in part due to the author's own lack of expertise in such matters, and in part due to the relative simplicity of skyrmion physics compared to other modern disciplines of condensed matter. Several edited volumes on skyrmion science have also been published in keeping with the rapid progress of the field. This book is intended to maintain a healthy distance from the "trendy" volumes by removing discussions of latest experimental results, which are thoroughly covered in other books and review articles. The manner of exposition is overtly pedagogical; I am convinced that skyrmions happen to be as good an excuse as any other physical platform in which to learn the modern tools of theoretical condensed matter physics.

Skyrmions have become ubiquitous in condensed matter physics. They can be found in a wide variety of magnets, in spinor cold atoms, and in liquid crystals. There are also momentum-space manifestations of skyrmions, as in two-dimensional topological band insulators, and in three-dimensional Weyl semi-metals. Here, however, I exclude most such examples, focusing on skyrmions in chiral magnets and cold atoms, finding it safer to talk about the themes where I seem to have genuine voice of confidence to share.

Suwon, Korea (Republic of) Jung Hoon Han
September 2017

Acknowledgements

This book was written at the suggestion of Professor Atsushi Fujimori during the workshop held in the Muju resort in Korea, 2012. I am very grateful to him for expressing his confidence in me as a writer.

I wish to thank all my teachers and collaborators over the years, especially my advisor David Thouless, postdoc supervisor Dung-Hai Lee, and my longtime mentor and collaborator Naoto Nagaosa. I am not ashamed to say that most of what I know in condensed matter physics came from working with them.

All the work I did on skyrmions came in the form of collaboration with various students and postdocs over the years. I thank (in chronological order of collaboration) Shigeki Onoda, Su Do Yi, Jin-Hong Park, Jiadong Zang, Maxim Mostovoy, Zhihua Yang, Xiao-Qiang Xu, Ye-Hua Liu, Hyunyong Lee, Yun-Tak Oh, and Seong-Gyu Yang, for the papers written and discussions held over the years. All the schematic figures in the book are the product of diligent work by Taehee Kim.

The completion of this book was delayed for a few years due to a number of unjustifiable reasons, one of which was a large grant I received that drowned me in postdocs and projects. When the announcement of Thouless' shared Nobel Prize came in October of 2016 (also when my grant was prematurely terminated), the urge to finish this book turned into an adrenaline rush.

A million thanks also go out to Manhyung and Jiwoo for steadily tunneling through their own high-school and middle-school careers in the absence of their mother. Their unflinching attitudes in times of crisis was an example to their father. As a book is valued more than a paper in my scholarly family atmosphere, this book is dedicated to my parents.

Contents

Acronyms

AB	Aharonov–Bohm
DM	Dzyaloshinskii–Moriya
GL	Ginzburg–Landau
GP	Gross–Pitaevskii
LL	Landau–Lifshitz
LLG	Landau–Lifshitz–Gilbert
MC	Monte Carlo
STT	Spin-transfer torque
$\mathbf{a}$	Emergent vector potential
$\mathbf{b}$	Emergent magnetic field
$\mathbf{e}$	Emergent electric field
q_e	Electric charge
q_v	Vortex charge
q_s	Skyrmion density
Q_s	Skyrmion number/charge
$\mathbf{q}$	Skyrmion density vector
∇_2	Two-dimensional gradient $(\partial_x, \partial_y, 0)$

Chapter 1
Geometric Phases

This is the first of two chapters covering the mathematical and theoretical preliminaries one needs to tackle the skyrmion problems. In particular, the materials of this chapter comprise fundamental aspects of the modern theory of magnetism. Nowadays, there exists a mostly uniform treatment of these topics across the many condensed matter theory books available. Despite the risk of banality, I have tried to include a self-contained chapter on the Berry phase for spin problems so that readers need not look elsewhere to garner the requisite background.

Starting with the well-known example of the Lagrangian for the motion of charged particles in magnetic fields, we show how a very similar geometric term should appear in the Lagrangian for the spin. The celebrated Landau–Lifshitz (LL) equation of magnetization dynamics then follows from the Euler–Lagrange equation arising from the spin–Lagrangian.

The so-called Berry phase action for spin may appear in a wide number of disguises. Among the ambitious goals of this chapter is to provide a careful presentation of all such disguises, and to finish by compiling a dictionary of all the equivalent expressions. Several mathematical identities discussed in this chapter will be useful throughout the remainder of this book.

1.1 Geometric Phase of a Charged Particle in a Magnetic Field

There are several names in concurrent use for the path-dependent, time-dependent piece of the Lagrangian. It is variously called: the geometric phase, the Berry phase, or the adiabatic phase, depending on one's taste and upbringing. In this book, we will use the first two terms interchangeably. The first (and presumably the most widely known) instance of the geometric phase is that of a charged particle with charge q_e

J.H. Han, *Skyrmions in Condensed Matter*, Springer Tracts in Modern Physics 278, https://doi.org/10.1007/978-3-319-69246-3_1

in a (nondynamic) background electromagnetic potential $(A_0, \mathbf{A})$. The Lagrangian of this system is then[1]

$$L = \frac{m}{2}\dot{\mathbf{r}}^2 + q_e(\dot{\mathbf{r}} \cdot \mathbf{A} - A_0), \tag{1.1}$$

where we note that, by assumption, the gauge field $(A_0, \mathbf{A})$ is considered to be both space- and time-dependent.

Variation of the action $S = \int dt\, L$ gives

$$\delta S = \int dt\, \delta\mathbf{r} \cdot \left(-m\ddot{\mathbf{r}} - q_e\dot{\mathbf{A}} - q_e\nabla A_0\right) + q_e[\dot{\mathbf{r}} \cdot \partial_\mu \mathbf{A}]\delta r_\mu, \tag{1.2}$$

where $\mu = x, y, z$, and the tricky part of the above calculation is the careful manner in which the time derivative $\dot{\mathbf{A}}$ must be handled:

$$\dot{\mathbf{A}} = \partial_t \mathbf{A} + (\dot{\mathbf{r}} \cdot \nabla)\mathbf{A}. \tag{1.3}$$

After some manipulation of the usual identities, $\mathbf{E} = -\nabla A_0 - \partial_t \mathbf{A}$ and $\mathbf{B} = \nabla \times \mathbf{A}$, the well-known Lorentz's equation may be derived from the stationary condition $\delta S = 0$:

$$m\ddot{\mathbf{r}} = q_e(\mathbf{E} + \dot{\mathbf{r}} \times \mathbf{B}). \tag{1.4}$$

The corresponding Hamiltonian then follows from the standard Legendre transformation

$$H = \mathbf{p} \cdot \dot{\mathbf{r}} - L = \frac{1}{2m}(\mathbf{p} - q_e\mathbf{A})^2 + q_e A_0, \tag{1.5}$$

where the canonical momentum is defined by $\mathbf{p} = \delta L/\delta\dot{\mathbf{r}} = m\dot{\mathbf{r}} + q_e\mathbf{A}$.

Having identified the correct Lagrangian, the action S, constructed as its time integral, may be shown to contain a piece that only depends on the path of the particle, since it can be cast in the form

$$S_{\mathrm{AB}} = q_e \int_{t_1}^{t_2} dt\, \dot{\mathbf{r}} \cdot \mathbf{A} = q_e \int_{\mathbf{r}_1}^{\mathbf{r}_2} d\mathbf{r} \cdot \mathbf{A}, \tag{1.6}$$

where the subscript AB stands for Aharonov–Bohm. A unique property of this path-dependent integral (extreme right of the above equation) is that it yields the same value regardless of how fast or slow the particle's motion is, so long as the paths followed by the particle are the same. The reason for calling this a "phase" is that

---

[1] We will not necessarily make a notational distinction between Lagrangian and its density; both are designated with L. More often than not, the spatial integral symbol will be missing in what I call the "Lagrangian".

the quantum–mechanical amplitude of the particle trajectory is given by $e^{iS/\hbar}$, which contains the geometric factor $e^{iS_{\mathrm{AB}}/\hbar}$.

Specializing to a constant magnetic field yields some interesting insights. For an arbitrary constant vector $\mathbf{B}$, the vector potential can be written as $\mathbf{A} = \mathbf{B} \times \mathbf{r}/2$, where $\mathbf{r} = (x, y, z)$ describes the position of the particle. The Lagrangian and Hamiltonian associated with this vector potential are,

$$L = \frac{m}{2}\dot{\mathbf{r}}^2 + \frac{q_e}{2}\mathbf{B} \cdot (\mathbf{r} \times \dot{\mathbf{r}}) - q_e A_0$$
$$H = \frac{1}{2m}\left(\mathbf{p} - \frac{q_e}{2}\mathbf{B} \times \mathbf{r}\right)^2 + q_e A_0, \tag{1.7}$$

respectively.

An interesting case of this theory is the infinite field $|\mathbf{B}| \to \infty$, or zero-mass $m \to 0$ limit. In this limit, the kinetic energy $(m/2)\dot{\mathbf{r}}^2$ drops out from the Lagrangian, but the geometric piece responsible for the AB phase survives. In particular, the AB phase round a closed path Γ, such as the path shown in Fig. 1.1, is equal to the area enclosed by the loop multiplied by the size of the perpendicular component of the field and the charge. This proportionality to the enclosed area is a manifestation of the geometric nature of the action.

From a dynamical point of view, in the massless limit, the left-hand side of the Lorentz equation of motion (1.4) reduces to $m\ddot{\mathbf{r}} = 0$, and we then find:

$$\dot{\mathbf{r}} \times \mathbf{B} = -\mathbf{E}. \tag{1.8}$$

When $\mathbf{E}$ and $\mathbf{B}$ are orthogonal, the equation can be solved to give $\dot{\mathbf{r}} = \mathbf{E} \times \mathbf{B}/(\mathbf{B} \cdot \mathbf{B})$. This expression describes the well-known guiding-center dynamics of charged particles, where the Newtonian dynamics picture of acceleration being proportional to the force gives way to a scenario where the motion is strictly orthogonal to the direction of force.

The guiding-center limit is thus not so far-fetched after all. Indeed, if the magnetic field is strong enough that the cyclotron energy $\hbar\omega_c = \hbar|q_e B|/m$ is much larger than any other energy scale affecting the electron dynamics, the motion of charges constrained within the lowest Landau level obeys the guiding-center dynamics given by Eq. (1.8). Moreover, as we will amply cover in Chap. 4, topological objects such as

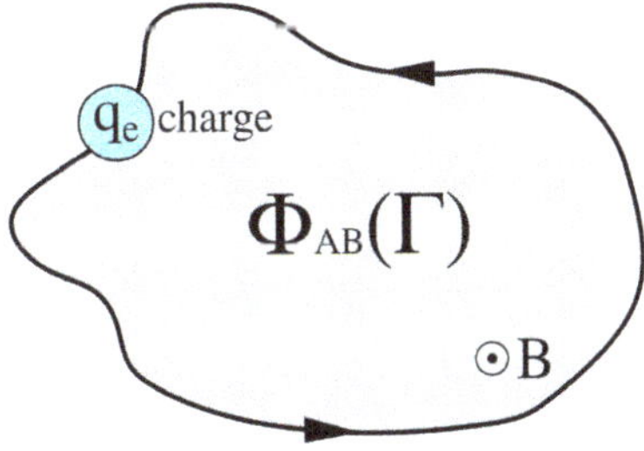

Fig. 1.1 Aharonov–Bohm phase factor $\Phi_{\mathrm{AB}}(\Gamma)$ round a closed path Γ is proportional to the area enclosed by it, provided the strength of magnetic field B perpendicular to it is constant

vortices in superfluids and skyrmions in magnets also obey guiding-center dynamics, despite the lack of a quantizing magnetic field to dictate their lowest Landau level physics. For now, we simply note that while the underlying microscopic mechanisms are vastly different, the emergent dynamics for all three examples are very similar, if not identical. The following statement is worth bearing in mind for the rest of the book:

$$\begin{aligned}&(\text{electrons in quantizing magnetic field})\\ &\approx (\text{vortices in superfluids})\\ &\approx (\text{skyrmions in magnets}).\end{aligned} \tag{1.9}$$

The $\approx$ symbol refers to the similarity in their respective point-particle dynamics.

1.2 Quantum Spin Dynamics

Most of this book is concerned with the dynamics of magnetic moments arising from electrons whose positions are localized in space and whose spins are free to rotate. The simplest illustration of spin dynamics comes from the Zeeman Hamiltonian, which is a quantum expression for the magnetic dipole energy $E_Z = -\boldsymbol{\mu} \cdot \mathbf{B}$. From classical electrodynamics, we know that the magnetic moment $\boldsymbol{\mu}$ of a charged particle in a circular orbit carrying angular momentum $\mathbf{L}$ is $\boldsymbol{\mu} = (q_e/2m)\mathbf{L}$. The electron spin also carries an angular momentum $\mathbf{S}$, but the associated magnetic moment has an extra proportionality factor g called the g-factor, which is approximately equal to 2. Putting everything together, one arrives at the Zeeman Hamiltonian:

$$H_Z = -\boldsymbol{\mu} \cdot \mathbf{B} = -g\frac{q_e}{2m_e}\mathbf{S} \cdot \mathbf{B}. \tag{1.10}$$

The notion of the Bohr magneton then arises when one writes the spin operator in terms of Pauli matrices, $\mathbf{S} = (\hbar/2)\boldsymbol{\sigma}$, and substitutes $q_e = -e$ and $g = 2$ above:

$$H_Z = \frac{e\hbar}{2m_e}\boldsymbol{\sigma} \cdot \mathbf{B} \equiv \mu_{\mathrm{B}}\boldsymbol{\sigma} \cdot \mathbf{B}. \tag{1.11}$$

The actual value of the Bohr magneton is then equal to $\mu_B = 5.7883817555(79) \times 10^{-5}$ eV T^{-1}, and leads to an energy separation between spin-up and spin-down electrons at $B = 10$ T of $\Delta E_Z \approx 1.1$ meV.

Physically, the electron spin executes a precessional motion when placed in a magnetic field, where the the precession axis is defined by the field direction. In order to see how this comes about, let us work out the equation of motion of the electron spin operator using the Heisenberg representation. The time evolution of a Heisenberg operator $\mathscr{A}(t) \equiv e^{iHt/\hbar}\mathscr{A}e^{-iHt/\hbar}$ follows directly from its commutator with the Hamiltonian, $\hbar\dot{\mathscr{A}}(t) = i[H, \mathscr{A}(t)]$. Applying the rule to the Zeeman Hamiltonian

(1.10) and using the commutation algebra of spin operators $[S_\alpha, S_\beta] = i\hbar\varepsilon_{\alpha\beta\gamma}S_\gamma$, one finds a simple first-order equation for the spin precession:

$$\dot{\mathbf{S}} = \mathbf{S} \times \left(-\frac{\delta H_Z}{\delta \mathbf{S}}\right). \tag{1.12}$$

This expression suggests that for a more general spin Hamiltonian H, the spin dynamics follows by replacing $\delta H_Z/\delta\mathbf{S}$ with the variational derivative $\delta H/\delta\mathbf{S}$.

Since in an experiment one measures the magnetic moment rather than the spin itself, it is better to convert (1.12) to an equation in terms of the magnetic moment $\boldsymbol{\mu}$. The linear relation between $\boldsymbol{\mu}$ and the spin $\mathbf{S}$ is, as discussed earlier,

$$\boldsymbol{\mu} = -\frac{ge}{2m}\mathbf{S} = -\gamma\mathbf{S}, \tag{1.13}$$

where the gyromagnetic ratio γ is defined as a positive quantity. In turn, one can view $\boldsymbol{\mu}$ as a classical vector of certain length, say M, and write $\boldsymbol{\mu} = M\mathbf{n}$, where $\mathbf{n}$ is a unit vector. Replacing $\mathbf{S}$ by $-M\mathbf{n}/\gamma$ in (1.12) gives

$$\dot{\mathbf{n}} = \frac{\gamma}{M}\mathbf{n} \times \frac{\delta H}{\delta \mathbf{n}}. \tag{1.14}$$

Through judicious choice of the overall scale, one can absorb M into the Hamiltonian H, and arrive at a simpler form of the equation

$$\dot{\mathbf{n}} = \gamma\mathbf{n} \times \frac{\delta H}{\delta \mathbf{n}}. \quad \text{(conventional LL)} \tag{1.15}$$

This is the semi-classical equation for spin dynamics known as the Landau–Lifshitz (LL) equation. The equation was introduced in 1935 by Landau and Lifshitz as a way to understand the dynamics of ferromagnets [1], and is widely used in the study of magnetization dynamics ever since. Although its origin can be traced to the quantum–mechanical spin algebra, the equation is treated as if the magnetic moments were classical vectors of fixed magnitude.

Let us analyze the LL equation (1.15) in some particular limits. For starters, one can readily prove that $\mathbf{n} \cdot \dot{\mathbf{n}} = 0$, since the right-hand side of (1.12) is orthogonal to $\mathbf{n}$. This result means that the magnetization dynamics only occurs in such a way as to conserve the magnitude of the magnetization at all times. One can, in fact, make the general argument that norm-preserving spin dynamics are constrained to obey an equation of the form $\dot{\mathbf{n}} = \mathbf{n} \times \mathbf{A} + \mathbf{n} \times (\mathbf{n} \times \mathbf{B})$ for any two model-dependent vectors $\mathbf{A}$ and $\mathbf{B}$, since the two terms on the right-hand side fully span the two directions orthogonal to $\mathbf{n}$ in three-dimensional space.

On decomposing the moment into its parallel ($\mathbf{n}_\parallel \parallel \delta H/\delta\mathbf{n}$) and orthogonal ($\mathbf{n}_\parallel \perp \delta H/\delta\mathbf{n}$) components with respect to the local field direction, $\mathbf{n} = \mathbf{n}_\parallel + \mathbf{n}_\perp$, the LL equation takes the form

$$\dot{\mathbf{n}}_{\parallel} + \dot{\mathbf{n}}_{\perp} = \gamma \mathbf{n}_{\perp} \times \frac{\delta H}{\delta \mathbf{n}}. \tag{1.16}$$

Since the cross product on the right-hand side yields a vector in the plane orthogonal to $\delta H/\delta \mathbf{n}$, we must have $\dot{\mathbf{n}}_{\parallel} = 0$. In the case of a constant effective magnetic field $\delta H/\delta \mathbf{n} = -\mathbf{B}_{\text{eff}}$, we may write:

$$\dot{\mathbf{n}}_{\perp} = \gamma \mathbf{B}_{\text{eff}} \times \mathbf{n}_{\perp}, \tag{1.17}$$

where the perpendicular component of the moment $\mathbf{n}_{\perp}$ executes a (right-handed) precessional motion with the angular frequency $\omega_{\text{L}} = \gamma |\mathbf{B}_{\text{eff}}|$. This angular frequency is known as the Larmor frequency around the magnetic field axis (see Fig. 1.2). The overall motion of the magnetic moment under a constant magnetic field is known as the Larmor precession.

Owing to the unfortunate historical incident of labeling the electronic charge as negative, the spin of an electron points in the opposite direction to the magnetization vector. Thus, the equation governing spin dynamics, as opposed to magnetization dynamics, is obtained by taking $\mathbf{n} \to -\mathbf{n}$ in the LL equation above,

$$\dot{\mathbf{n}} = -\gamma \mathbf{n} \times \frac{\delta H}{\delta \mathbf{n}}. \quad \text{(our LL)} \tag{1.18}$$

This is the form of the LL equation adopted for the remainder of the book. Readers who may be confused by the different sign conventions of the LL equation across the literature should simply remind themselves that some authors prefer to think in terms of spin rather than the magnetization vector.

Now that we know the equation of motion for spin is the LL equation, it is fitting to ask whether there exists a Lagrangian whose variational derivative can reproduce

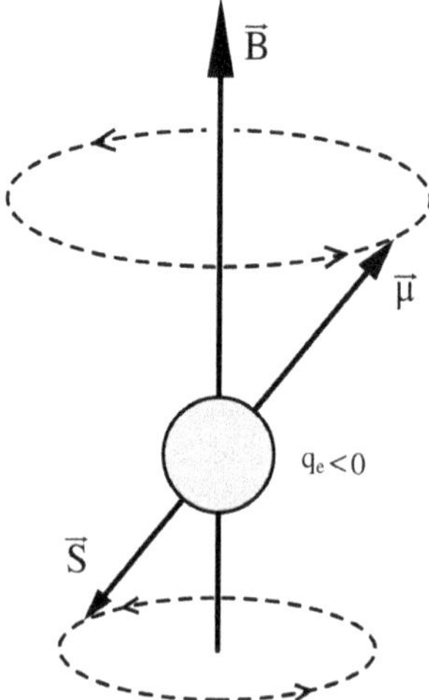

Fig. 1.2 Larmor precession of the magnetic moment $\boldsymbol{\mu}$ about the magnetic field **B**. For negatively charged particles, the spin **S** points opposite to the magnetic moment

it. The route from the Hamiltonian to the Lagrangian in classical mechanics is to first identify the canonical pair (e.g., $\mathbf{p}$ and $\mathbf{r}$), and to then employ the Legendre transformation $L = \mathbf{p} \cdot \dot{\mathbf{r}} - H$, using the first of Hamilton's equations $\partial H/\partial \mathbf{p} = \dot{\mathbf{r}}$ to rewrite $\mathbf{p}$ in terms of $\dot{\mathbf{r}}$ and $\mathbf{r}$. However, this approach does not work for spins because the spin Hamiltonian depends on two coordinates, i.e., the polar angle θ and the azimuthal angle ϕ of the spin orientation, rather than on one coordinate and its time derivative. We will shortly learn that the two coordinates do actually form a conjugate pair, but at this stage, this fact is not obvious.

A proper quantum-mechanical derivation of the spin Lagrangian will be given in the next section using the coherent state formalism. For now, we content ourselves by simply writing down the answer and showing that it does reproduce the LL equation. The Lagrangian is:

$$L = L_{\mathrm{B}} - \hbar\gamma H = -\hbar S(1 - \cos\theta)\dot{\phi} - \hbar S\gamma H. \tag{1.19}$$

Here, the subscript B stands for M. V. Berry, who wrote down the spin version of the AB phase in 1984 [2]. In honor of his discovery, nowadays, this is often called the Berry phase. In order to distinguish it from geometric phase factors arising in other contexts, we will also refer to it as the Berry phase for spin. Here, the prefactor $\hbar S\gamma$ appears before the Hamiltonian H so as to match the symbol H appearing in the LL equation.

In general, the unit vector $\mathbf{n} = (\sin\theta\cos\phi, \sin\theta\sin\phi, \cos\theta)$ is used to define the coherent-state counterpart of the spin-S operator $\mathbf{S} = \hbar S\mathbf{n}$, and H is some function of $\mathbf{n}$. We then first keep in mind the following two results for the derivatives of the unit vector $\mathbf{n}$ along the two tangential directions:

$$\begin{aligned} \frac{\partial \mathbf{n}}{\partial \theta} &= (\cos\theta\cos\phi, \cos\theta\sin\phi, -\sin\theta) = \hat{\theta}, \\ \frac{\partial \mathbf{n}}{\partial \phi} &= (-\sin\phi, \cos\phi, 0)\sin\theta = \hat{\phi}\sin\theta. \end{aligned} \tag{1.20}$$

Variation of the action $S = \int dt\, L$ gives a pair of equations:

$$\begin{aligned} -\hbar S\dot{\mathbf{n}} \cdot \hat{\phi} - \hbar S\gamma\frac{\delta H}{\delta \mathbf{n}} \cdot \hat{\theta} &= 0, \\ \hbar S\dot{\mathbf{n}} \cdot \hat{\theta} - \hbar S\gamma\frac{\delta H}{\delta \mathbf{n}} \cdot \hat{\phi} &= 0, \end{aligned} \tag{1.21}$$

where the two time derivatives $\dot{\theta}$ and $\dot{\phi}$ arising from the variation of the action have been replaced by

$$\dot{\mathbf{n}} \cdot \hat{\theta} = \dot{\theta}, \quad \dot{\mathbf{n}} \cdot \hat{\phi} = \dot{\phi}\sin\theta. \tag{1.22}$$

Secondly, the mutual orthogonality of the three unit vectors $\mathbf{n}, \hat{\theta}, \hat{\phi}$ allows us to write $\hat{\theta} = \hat{\phi} \times \mathbf{n}$ and $\hat{\phi} = \mathbf{n} \times \hat{\theta}$. The two Euler-Lagrange equations of (1.21) are thus re-cast in the form

$$\hbar S \left(-\dot{\mathbf{n}} + \gamma \frac{\delta H}{\delta \mathbf{n}} \times \mathbf{n} \right) \cdot \hat{\phi} = 0,$$
$$\hbar S \left(\dot{\mathbf{n}} - \gamma \frac{\delta H}{\delta \mathbf{n}} \times \mathbf{n} \right) \cdot \hat{\theta} = 0. \tag{1.23}$$

These two equations are nothing but the projections of the LL equation (1.18) along the two tangential directions.

It is instructive to think about the similarity of the spin action to that of a charged particle moving in a magnetic field. For one, the mass term proportional to $\dot{\mathbf{n}}^2$ must be absent in the spin action, since otherwise the length conservation $\mathbf{n} \cdot \dot{\mathbf{n}} = 0$ would be destroyed in the Euler–Lagrange equation. In order to preserve the norm of the spin, therefore, mass-like terms simply cannot appear in the action. With respect to particle motion, the absence of mass led to the strict orthogonality of the motion to the external force, resulting in the guiding-center dynamics governing the motion [see Eq. (1.8)]. Moreover, the part of Lagrangian that survived in the massless limit for the charged particle was the AB phase. We may then ask if it is fair to interpret the geometric phase of spin as a sort of AB phase. The answer is in fact yes, as the following argument taken from Witten's 1983 landmark paper [3] shows.[2]

Imagine a particle of zero mass that is constrained to move on the surface of a unit sphere, as shown in Fig. 1.3. Its coordinates are given by $\mathbf{n}$, and its velocity is given by $\dot{\mathbf{n}} = \hat{\theta}\dot{\theta} + \hat{\phi}\dot{\phi} \sin\theta$. As in our earlier discussion regarding the action of a charged particle subject to electromagnetic fields, there will be an AB component in the action

$$L_{\text{AB}} = q_e \mathbf{A} \cdot \dot{\mathbf{n}} = q_e(\mathbf{A} \cdot \hat{\theta})\dot{\theta} + q_e(\mathbf{A} \cdot \hat{\phi})\dot{\phi} \sin\theta. \tag{1.24}$$

Suppose now that this Lagrangian could be identified with the Berry phase, $L_{\text{B}} = -\hbar S(1 - \cos\theta)\dot{\phi}$. In order to make the equality $L_{\text{AB}} = L_{\text{B}}$ work, the following equations need to hold:

$$q_e \mathbf{A} \cdot \hat{\theta} = 0$$
$$q_e \mathbf{A} \cdot \hat{\phi} = \hbar S \frac{\cos\theta - 1}{\sin\theta}. \tag{1.25}$$

[2] It is an interesting coincidence that Witten's 1983 paper basically dealt with the Berry's phase term, or the topological term, in field theory. A year before Witten, the Thouless-Kohmoto-Nightingale-denNijs (TKNN) paper on the derivation of the topological number for the integer quantum Hall effect made essential use of the idea of geometric connection [4]. Berry's own work, which came to stand for many (if not most) of geometric effects in condensed matter physics, was published in 1984 [2]. In retrospect, the early 80 s might go down as an era of the massive infusion of topological ideas in theoretical physics. Of course, Dirac, Aharonov, and Bohm had laid the ground a long time before.

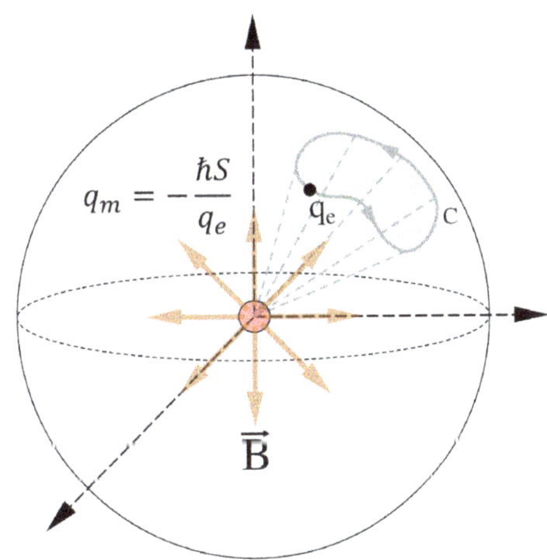

Fig. 1.3 The spin Lagrangian is similar to the Aharonov-Bohm phase of a massless particle moving on a unit sphere subject to a magnetic field arising from a magnetic monopole located at the center of the sphere

One such vector potential satisfying these conditions is

$$\mathbf{A} = \frac{\hbar S}{q_e} \frac{\cos\theta - 1}{\sin\theta} \hat{\phi}. \tag{1.26}$$

If this vector function should elicit a sense of *déjà vu*, that is because we learned that the vector potential of a magnetic monopole looks just like that:

$$\mathbf{A} = \frac{\hbar S}{q_e r} \frac{\cos\theta - 1}{\sin\theta} \hat{\phi} \Rightarrow \nabla \times \mathbf{A} = -\frac{\hbar S}{q_e} \frac{\hat{r}}{r^2}. \tag{1.27}$$

In other words, the geometric phase for spin is exactly that of a charged particle of unit charge $q_e = +1$, constrained to move on the surface of the unit sphere in the presence of a magnetic field arising from a magnetic monopole of charge $q_m = -\hbar S/q_e$ located at the center of the sphere. The flux through the unit sphere is, therefore, $\Phi = -4\pi\hbar S/q_e = -2hS/q_e$. Let us suppose that the flux quantization rule applied here as well; then, Φ would have to be an integer multiple of the flux quantum, $\Phi = (h/q_e) \times n$. In other words, $2S = n$ would have to be an integer, which expresses the familiar half-integer quantization of the spin S.

The analogy to the AB phase also suggests the relation

$$\oint_C dt\ (1 - \cos\theta)\partial_t \phi = A_C, \tag{1.28}$$

for the spin Lagrangian, which expresses the fact that the integral of the geometric phase along a closed path is proportional to the area enclosed by the loop on the unit sphere (shaded area in Fig. 1.3).

1.3 Coherent-State Path Integral for Spins

We now ask how we may derive the Lagrangian (1.19) along the lines of Feynman's path integral approach. With this approach, one deals with quantum-mechanical averages of operators, which are numbers that can be varied continuously, rather than operators, which cannot. This is easy to imagine for averages of operators like the momentum, $\mathbf{p}$, because we know that $\langle\mathbf{k}|\mathbf{p}|\mathbf{k}\rangle = \mathbf{k}$ can be any vector we like, provided the plane-wave state $|\mathbf{k}\rangle$ is defined in an infinitely large medium. However, for spins whose eigenvalues are discrete quantized values ranging from $+S$ to $-S$ in integer steps, it is less clear how to integrate over intermediate states, or how to think about smooth variations of the possible paths.

Fortunately, there is a neat way to define spin states in such a manner that the average of the spin operator turns out to be a continuously varying vector, just as if it were a classical quantity. In such a basis, one can formulate the path integral of the spin system in much the same way as was done for the ordinary particle motion. The underlying method is called the spin coherent state.[3]

Let us introduce the idea in the context of spin-S quantum mechanics. First, the state corresponding to the largest expectation value (average) of the S^z operator, i.e., the z-component of the spin operator, is denoted by $|\mathbf{n}_0, S\rangle$, and assigned the column vector representation $\begin{pmatrix}1 & 0 & \cdots & 0\end{pmatrix}^T$. This assignment is possible because we are using a diagonal representation for the spin operator S^z, i.e., $S^z = \hbar\,\mathrm{diag}(S, S-1, \cdots, -S+1, -S)$. The representation of the other spin components is then fixed by our choice of S^z as well as the commutation relation relating the spin components $[S_\alpha, S_\beta] = i\hbar\varepsilon_{\alpha\beta\gamma}S_\gamma$, which are valid for arbitrary spin-S. It is easy to check that the average of the spin operator for this state is $\langle\mathbf{n}_0, S|\mathbf{S}|\mathbf{n}_0, S\rangle = S\mathbf{n}_0$, where $\mathbf{n}_0 = (0, 0, 1)$ defines the north direction of the unit sphere. Suppose now that one wished to write down a spinor state $|\mathbf{n}, S\rangle$ such that its average spin orientation is along some arbitrary direction $\mathbf{n}$ on the unit sphere. It would be sufficient to find a state that satisfied the eigenvalue problem

$$\mathbf{S}\cdot\mathbf{n}|\mathbf{n}, S\rangle = \hbar S|\mathbf{n}, S\rangle, \tag{1.29}$$

since this would immediately imply $\langle\mathbf{n}, S|\mathbf{S}\cdot\mathbf{n}|\mathbf{n}, S\rangle = \hbar S$. This state $|\mathbf{n}, S\rangle$ is called the spin-S coherent state.

In classical mechanics, one way to construct an arbitrary vector $\mathbf{n}$ on the unit sphere is through an Euler rotation of the reference vector $\mathbf{n}_0$. Specifically, an Euler rotation can be written as the product of three rotations as

$$\mathscr{R}(\alpha, \beta, \gamma) = \mathscr{R}_z(\alpha)\mathscr{R}_y(\beta)\mathscr{R}_z(\gamma), \tag{1.30}$$

where each $\mathscr{R}_\alpha(\varphi_\alpha)$ is a SO(3) matrix describing a rotation about the α-axis by the angle φ_α. Each rotation matrix is also related to another matrix J_α, called the genera-

[3]The word "coherent state" appears in many other contexts of quantum physics. Here, for clarity, we use the term "spin coherent state".

tor, by an exponential relation: $\mathscr{R}_\alpha(\varphi_\alpha) = \exp(-i\varphi_\alpha J_\alpha)$. Since the SO(3) matrices are real (no imaginary numbers in classical mechanics!), the generators must be purely imaginary and anti-symmetric: $[J_\alpha]_{\beta\gamma} = -i\varepsilon_{\alpha\beta\gamma}$. Owing to our choice of representation, the first matrix $\mathscr{R}_z(\gamma)$ acting on $\mathbf{n}_0$ does not produce a change, i.e., $\mathscr{R}_z(\gamma)\mathbf{n}_0 = \mathbf{n}_0$, while the other two rotations bring $\mathbf{n}_0$ to $\mathbf{n} = (\sin\beta\cos\alpha, \sin\beta\sin\alpha, \cos\beta)$, where the polar and azimuthal angles are given by β and α, respectively.

On the other hand, rotations of spinor wave functions are carried out by SU(2) unitary matrices. Their generators, which must be Hermitian in order to meet unitarity requirements, are the spin operators themselves, $J_\alpha = S_\alpha/\hbar$, with matrix dimensions $(2S+1)\times(2S+1)$. The spin-S coherent state may thus be obtained through an Euler rotation of the highest-weight state $|\mathbf{n}_0, S\rangle$:

$$|\mathbf{n}, S\rangle = e^{-i\alpha S^z} e^{-i\beta S^y} e^{-i\gamma S^z} |\mathbf{n}_0, S\rangle. \tag{1.31}$$

Since $|\mathbf{n}_0, S\rangle$, by definition, is an eigenstate of S^z with $S^z|\mathbf{n}_0, S\rangle = S|\mathbf{n}_0, S\rangle$, one can express the coherent state with only two operators:

$$|\mathbf{n}, S\rangle = e^{-i\gamma S} e^{-i\alpha S^z} e^{-i\beta S^y} |\mathbf{n}_0, S\rangle. \tag{1.32}$$

We recall that the proof one needs to establish is then

$$\mathbf{S}\cdot\mathbf{n}\Big(e^{-i\alpha S^z} e^{-i\beta S^y} |\mathbf{n}_0, S\rangle\Big) = \hbar S\Big(e^{-i\alpha S^z} e^{-i\beta S^y} |\mathbf{n}_0, S\rangle\Big). \tag{1.33}$$

It is useful to invoke the following mathematical identity ($a \neq b$)

$$e^{i\theta S_a} S_b e^{-i\theta S_a} = S_b \cos\theta - \varepsilon_{abc} S_c \sin\theta, \tag{1.34}$$

which is valid for arbitrary spin S. With this identity, one can first show that

$$\begin{aligned} e^{i\alpha S^z}(\mathbf{S}\cdot\mathbf{n})e^{-i\alpha S^z} &= (S^x\cos\alpha - S^y\sin\alpha)n^x + (S^y\cos\alpha + S^x\sin\alpha)n^y + S^z n^z \\ &= S^x\sin\beta + S^z\cos\beta. \end{aligned} \tag{1.35}$$

A second unitary transformation then gives

$$e^{i\beta S^y}\big(S^x\sin\beta + S^z\cos\beta\big)e^{-i\beta S^y} = S^z. \tag{1.36}$$

Thus, since $S^z|\mathbf{n}_0, S\rangle = \hbar S|\mathbf{n}_0, S\rangle$, one has established the desired relation (1.33) for arbitrary spin S.

Although the general proof is already established, one can still learn something from examining particular cases for different spin, starting with the most familiar one, $S = 1/2$, where the generators may be expressed in terms of the Pauli matrices $\mathbf{J} = \boldsymbol{\sigma}/2$. The spin coherent state is readily derived:

$$e^{-i(\alpha/2)\sigma^z}e^{-i(\beta/2)\sigma^z}e^{-i(\gamma/2)\sigma^z}\begin{pmatrix}1\\0\end{pmatrix}=e^{-i\gamma/2}\begin{pmatrix}e^{-i\alpha/2}\cos[\beta/2]\\ e^{i\alpha/2}\sin[\beta/2]\end{pmatrix}\equiv\left|\mathbf{n},\frac{1}{2}\right\rangle. \tag{1.37}$$

With a little more exercise one can derive the $S=1$ coherent state:

$$|\mathbf{n},1\rangle=\begin{pmatrix}e^{-i\alpha}\cos^2(\beta/2)\\ (1/\sqrt{2})\sin\beta\\ e^{i\alpha}\sin^2(\beta/2)\end{pmatrix},\quad (\mathbf{S}\cdot\mathbf{n})|\mathbf{n},1\rangle=\hbar|\mathbf{n},1\rangle. \tag{1.38}$$

A final mathematical ingredient needed in the path-integral formulation is the existence of a complete set that can be inserted at every intermediate time interval. Using coherent states, this complete set is offered by the integral

$$\frac{1}{2\pi}\int_0^\pi \sin\beta d\beta\int_0^{2\pi}d\alpha|\mathbf{n},S\rangle\langle\mathbf{n},S|\equiv\int d\mathbf{n}|\mathbf{n},S\rangle\langle\mathbf{n},S|=1_{(2S+1)\times(2S+1)}. \tag{1.39}$$

It is worth mentioning that although different position or momentum states are orthogonal, $\langle\mathbf{r}'|\mathbf{r}\rangle=0=\langle\mathbf{p}'|\mathbf{p}\rangle$, two spin coherent states are not necessarily so: $\langle\mathbf{n}'|\mathbf{n}\rangle\neq 0$. For this reason, the set of all possible **n** orientations is said to form an over-complete set. However, for the purpose of evaluating the matrix element appearing in the path integral, all one needs is the identity property given by Eq. (1.39), and we can use the over-complete set for as long as it suits us. Finally, we note that the integration measure used in the complete set construction above is the areal element of the unit sphere.

The main item of interest in the path-integral construction is to calculate the amplitude

$$\langle\mathbf{n}_j,\tau_j|e^{-\Delta\tau\ H/\hbar}|\mathbf{n}_i,\tau_i\rangle\approx\langle\mathbf{n}_j,\tau_j|\mathbf{n}_i,\tau_i\rangle-\frac{\Delta\tau}{\hbar}\langle\mathbf{n}_j,\tau_j|H|\mathbf{n}_i,\tau_i\rangle, \tag{1.40}$$

where the imaginary-time formalism is assumed. The first factor appearing on the right-hand side is the overlap of the spin coherent states at slightly different times. Assuming a "smooth" variation of the spin's trajectory with time, one can Taylor-expand $|\mathbf{n}_j,\tau_j\rangle\approx|\mathbf{n}_i,\tau_i\rangle+\Delta\tau|\partial_\tau\mathbf{n}_i,\tau_i\rangle$ and write

$$\langle\mathbf{n}_j,\tau_j|e^{-\Delta\tau\ H}|\mathbf{n}_i,\tau_i\rangle\approx 1-\frac{\Delta\tau}{\hbar}\Big(-\hbar\langle\partial_\tau\mathbf{n}_i,\tau_i|\mathbf{n}_i,\tau_i\rangle+\langle\mathbf{n}_i,\tau_i|H|\mathbf{n}_i,\tau_i\rangle\Big). \tag{1.41}$$

Then, since $\langle\mathbf{n}|\mathbf{S}|\mathbf{n}\rangle=\hbar S\mathbf{n}$ and H is a function of $\mathbf{S}$, we can write

$$\langle\mathbf{n}_i,\tau_i|H|\mathbf{n}_i,\tau_i\rangle=H(\hbar S\mathbf{n}(\tau_i))=H, \tag{1.42}$$

which is a complex number.

Exponenting (1.41) yields the the propagator over an infinitesimal time interval

$$U(\mathbf{n}_j, \tau_j; \mathbf{n}_i, \tau_i) = \exp\Big(-\frac{\Delta\tau}{\hbar}\left[-\hbar\langle\dot{\mathbf{n}}(\tau_i)|\mathbf{n}(\tau_i)\rangle + H\right]\Big), \tag{1.43}$$

and evaluation of the propagator over a slightly different time segment leads inevitably to the appearance of a quantity called the "connection". For the spin problem, the connection is $\langle\dot{\mathbf{n}}|\mathbf{n}\rangle = -\langle\mathbf{n}|\dot{\mathbf{n}}\rangle$, which measures how two spin coherent states at slightly different times overlap. The long-time propagator for spin-S is given by the path integral

$$\begin{aligned} U(\mathbf{n}_f, \tau_f; \mathbf{n}_i, \tau_i) &= \int \mathscr{D}\mathbf{n}\exp\Big(-\frac{1}{\hbar}\int d\tau\big[\hbar\langle\mathbf{n}|\partial_\tau\mathbf{n}\rangle + H[\mathbf{n}]\big]\Big) \\ &\to \int \mathscr{D}\mathbf{n}\exp\Big(\frac{i}{\hbar}\int dt\big[i\hbar\langle\mathbf{n}|\partial_t\mathbf{n}\rangle - H[\mathbf{n}]\big]\Big), \end{aligned} \tag{1.44}$$

where the second line defines the propagator in the real time, $t = -i\tau$.

Using the definition of $|\mathbf{n}\rangle$ and the identity (1.34), it is then easy to derive the result[4]

$$\langle\mathbf{n}|\partial_t\mathbf{n}\rangle = iS(1-\cos\theta)\partial_t\phi. \tag{1.45}$$

The Berry phase expression thus remains identical under a transformation from imaginary time to the real time. To conclude, we have that the spin Lagrangian one obtains from the path integral formulation is

$$L = i\hbar\langle\mathbf{n}|\dot{\mathbf{n}}\rangle - H = \hbar S(\cos\theta - 1)\dot{\phi} - H. \tag{1.46}$$

1.4 Alternative Ways to Write the Geometric Phase

In the previous section, the derivation of the path-integral action for spin was carried out successfully for a particle with an arbitrary spin S. In this section, we will explore other ways of writing the geometric action, and by the end of this section one will be surprised (and delighted) by the many different ways that the same action can be expressed!

[4]We revert to using (θ, ϕ) to designate the angles of the unit vector.

1.4.1 CP^1 Action

Let us first introduce a new terminology called the CP^1 representation and write

$$\mathbf{z} = \begin{pmatrix} \cos\theta/2 \\ e^{i\phi}\sin\theta/2 \end{pmatrix}, \quad \mathbf{z}^\dagger \boldsymbol{\sigma} \mathbf{z} = \mathbf{n}. \tag{1.47}$$

Readers mat have realized that the CP^1 representation is simply another name for the spin-1/2 coherent state $|\mathbf{n}, 1/2\rangle$, and that the expression $\mathbf{z}^\dagger\boldsymbol{\sigma}\mathbf{z} = \mathbf{n}$ is an alternative way of defining the spin coherent-state average, $\langle\mathbf{n}, 1/2|\mathbf{S}|\mathbf{n}, 1/2\rangle = (\hbar/2)\mathbf{n}$. After all, the very derivation of the geometric spin action was based on the CP^1 representation. Crucially, the identity $-i\mathbf{z}^\dagger\partial_t\mathbf{z} = \partial_t\phi[1 - \cos\theta]/2$ allows us to write the geometric phase as

$$\exp\left(iS\int dt(\cos\theta - 1)\partial_t\phi\right) = \exp\left(2iS\int dt\ [i\mathbf{z}^\dagger\partial_t\mathbf{z}]\right). \tag{1.48}$$

Since $\mathbf{n}$ follows from $\mathbf{z}$ through $\mathbf{n} = \mathbf{z}^\dagger\boldsymbol{\sigma}\mathbf{z}$, the entire spin Lagrangian (1.46) can afford a consistent CP^1 description.

However, a mere re-writing of the spin action would not be a sufficient justification for introducing the new variable $\mathbf{z}$. It turns out that $\mathbf{z}$ has a number of special properties that makes the CP^1 description a powerful tool for the analysis of spin dynamics. Firstly, one may use $\mathbf{z}$ to define a gauge field,

$$a_\mu = -i\mathbf{z}^\dagger\partial_\mu\mathbf{z} = \frac{1}{2}(1 - \cos\theta)\partial_\mu\phi, \tag{1.49}$$

for any spacetime coordinate μ. Using this gauge field, one can prove a nice identity[5]

$$\begin{aligned}\frac{1}{2}\mathbf{n}\cdot(\partial_\mu\mathbf{n}\times\partial_\nu\mathbf{n}) &= \frac{1}{2}\sin\theta[(\partial_\mu\theta)(\partial_\nu\phi) - (\partial_\nu\theta)(\partial_\mu\phi)] \\ &= \partial_\mu a_\nu - \partial_\nu a_\mu.\end{aligned} \tag{1.50}$$

Therefore, for an arbitrary pair of spacetime coordinates μ and ν, the field tensor $f_{\mu\nu} = \partial_\mu a_\nu - \partial_\nu a_\mu$ equals the expression for the $\mathbf{n}$-vector shown on the left-hand side of the above equation, and which is known as the "topological density" or "skyrmion density".

We should also highlight an intriguing analogy to electrodynamics here. Focusing on the spatial components of the above relation, one finds that the quantity

$$\frac{1}{2}\mathbf{n}\cdot(\partial_\mu\mathbf{n}\times\partial_\nu\mathbf{n}) = \varepsilon_{\mu\nu\lambda}b_\lambda \tag{1.51}$$

---

[5] In the context of superfluid ^{3}He this identity is known as the Mermin-Ho relation.

plays the role of an "emergent" magnetic field b_λ, while

$$\frac{1}{2}\mathbf{n}\cdot(\partial_t\mathbf{n}\times\partial_\mu\mathbf{n}) = \partial_t a_\mu - \partial_\mu a_t = e_\mu \tag{1.52}$$

plays the role of an emergent electric field. Such electrodynamics analogy will play a pivotal role when we come to discuss skyrmion dynamics in later chapters.

1.4.2 Wess-Zumino Action

Now, in a highly abstract move, let us introduce one more coordinate u and a fictitious vector potential a_u along this new coordinate's direction. The unit vector $\mathbf{n}(t,u)$ is, accordingly, a function of both coordinates, t and u. On the other hand, since the new coordinate u is a mathematical artifact, one can endow it with any property one deems useful. For our purpose, we will choose u such that at $u=0$ the vector $\mathbf{n}(t,u)$ points along $\mathbf{n}(t,0)=\mathbf{n}_0=(0,0,1)$, regardless of the value of t, while at $u=1$, it points along the "physical" spin direction $\mathbf{n}(t,u=1)=\mathbf{n}(t)$. One can then visualize the situation where, for every spin vector $\mathbf{n}(t)$ at time t, we have a trajectory of vectors $\mathbf{n}(t,u)$ that start from $\mathbf{n}_0$ at $u=0$ and move along the great circle of the unit sphere to reach $\mathbf{n}(t)$ at $u=1$. This kind of dependence is easily realized by generalizing $\theta(t)$ to $\theta(t,u)=u\theta(t)$ so that one has

$$\begin{aligned}\theta(t,0) &= 0 \quad \text{(north pole)}\\ \theta(t,1) &= \theta(t) \quad \text{(physical spin)}.\end{aligned} \tag{1.53}$$

We note that since by construction the trajectory is assumed to trace out a path along a great circle of the unit sphere, we do not need to generalize the definition of $\phi(t)$, i.e., $\phi(t,u)=\phi(t)$, and therefore a_u will vanish: $(1-\cos\theta)\partial_u\phi/2=0$. The mathematical identity (1.50) in this situation becomes

$$\frac{1}{2}\mathbf{n}\cdot(\partial_t\mathbf{n}\times\partial_u\mathbf{n}) = \partial_t a_u - \partial_u a_t = -\partial_u a_t. \tag{1.54}$$

This is a first-order differential equation for a_t, and can be readily integrated to yield the "physical" (i.e., valid at $u=1$) gauge field $a_t(t)$:

$$a_t(t,u=1) \equiv a_t(t) = -\frac{1}{2}\int_0^1 du\, \mathbf{n}\cdot(\partial_t\mathbf{n}\times\partial_u\mathbf{n}). \tag{1.55}$$

Moreover, its value at $u=0$ is zero, since $\theta(t,0)=0$ and thus $a_t(t,0)=(1-\cos 0)\partial_t\phi/2=0$. As a result, the geometric phase, formerly written as a single integral of the gauge field $a_t(t)$ over time t, now becomes a double integral:

$$\exp\left(-2iS\int dt\, a_t(t)\right) = \exp\left(iS\int dt \int_0^1 du\ \mathbf{n}\cdot(\partial_t\mathbf{n}\times\partial_u\mathbf{n})\right)$$
$$= \exp(iS_{\text{WZ}}/\hbar). \tag{1.56}$$

We have managed to express geometric action entirely in terms of the **n**-vector, but at the expense of introducing a fictitious direction u. The geometric action written in a space with one extra dimension with the aid of a fictitious variable in this manner is commonly known as Wess–Zumino action in the literature.[6]

Since the Wess–Zumino action is quite a novel concept, some care is needed in order to figure out how to take its variation δS_{WZ}. First, there needs to be a scheme to implement the variation $\mathbf{n}(t)\to\mathbf{n}(t)+\delta\mathbf{n}(t)$ in terms of the fictitious field $\mathbf{n}(t,u)$. Again, this scheme can be devised based on mathematical convenience. For example, instead of letting u vary from 0 to 1, we let it vary from 0 to $1+\varepsilon$, and let $\mathbf{n}(t,1)=\mathbf{n}(t)$, and $\mathbf{n}(t,1+\varepsilon)=\mathbf{n}(t)+\delta\mathbf{n}(t)$. Then, a straightforward calculation

$$\begin{aligned}
\delta S_{\text{WZ}} &= \hbar S\int dt\int_1^{1+\varepsilon} du\ \mathbf{n}(t,1)\cdot[\partial_t\mathbf{n}(t,1)\times\partial_u\mathbf{n}(t,1)]\\
&= \hbar S\int dt\ \mathbf{n}(t,1)\cdot[\partial_t\mathbf{n}(t,1)\times(\partial_u\mathbf{n}(t,1)\cdot\varepsilon)]\\
&= \hbar S\int dt\ \mathbf{n}(t)\cdot[\partial_t\mathbf{n}(t)\times\delta\mathbf{n}(t)]\\
&= \hbar S\int dt\ [\mathbf{n}(t)\times\partial_t\mathbf{n}(t)]\cdot\delta\mathbf{n}(t)
\end{aligned} \tag{1.57}$$

leads to the equation of motion

$$\hbar S\mathbf{n}\times\partial_t\mathbf{n} - \hbar S\gamma\frac{\delta H}{\delta\mathbf{n}} = 0 \ \to\ \dot{\mathbf{n}} = \gamma\frac{\delta H}{\delta\mathbf{n}}\times\mathbf{n}, \tag{1.58}$$

which is entirely equivalent to our prior derivation, (1.18).

1.4.3 Magnetic Monopole Action

We have just learned that the geometric phase has an interesting way of being written in terms of the Wess–Zumino form, and that its variation equals

$$\frac{\delta S_{\text{WZ}}}{\delta\mathbf{n}} = \hbar S\mathbf{n}\times\partial_t\mathbf{n}. \tag{1.59}$$

[6]Competing nomenclatures include Wess–Zumino–Witten action, and the Wess–Zumino–Novikov–Witten action. For convenience, we will simply call it the Wess–Zumino action.

It was also mentioned earlier that the geometric phase may be interpreted as the AB phase of a charged particle on the sphere subject to a field arising from a magnetic monopole. We will explore this statement a little deeper here, because a lot of the literature treats the geometric phase in this way, and it is one of the most common ways to understand the Berry phase.

In (1.46), we saw that the geometric part of the spin action could be written

$$L_{\rm B} = i\hbar\langle\mathbf{n}|\dot{\mathbf{n}}\rangle = i\hbar\dot{\mathbf{n}}\cdot\langle\mathbf{n}|\boldsymbol{\nabla}_{\mathbf{n}}|\mathbf{n}\rangle. \tag{1.60}$$

Writing $\mathbf{A}[\mathbf{n}] = i\hbar\langle|\mathbf{n}|\boldsymbol{\nabla}_{\mathbf{n}}|\mathbf{n}\rangle$, the Berry phase term can also be written as an AB factor

$$L_{\rm B} = \dot{\mathbf{n}}\cdot\mathbf{A}[\mathbf{n}] = \hbar S(\cos\theta - 1)\dot{\phi}. \tag{1.61}$$

We may choose the gauge such that $\mathbf{A} = A_{\phi}\hat{\phi}$ with $A_{\phi} = \hbar S(\cos\theta - 1)/\sin\theta$, i.e.,

$$\mathbf{A} = \hbar S\frac{\cos\theta - 1}{\sin\theta}\hat{\phi} = -\hbar S\frac{\sin\theta}{1+\cos\theta}\hat{\phi}, \tag{1.62}$$

in which case the singular line (Dirac string) extends along $\theta = \pi$. A more general choice of the vector potential is

$$\mathbf{A}[\mathbf{n}] = -\hbar S\frac{\mathbf{n}_0\times\mathbf{n}}{1-\mathbf{n}_0\cdot\mathbf{n}}, \tag{1.63}$$

for which the Dirac string is extended along the arbitrary direction $\mathbf{n}_0$. Regardless of the choice of $\mathbf{n}_0$, taking the curl always yields the same magnetic field

$$\boldsymbol{\nabla}_{\mathbf{n}}\times\mathbf{A} = \hbar S\mathbf{n}. \tag{1.64}$$

Readers can show this explicitly for a simple choice of $\mathbf{n}_0$ such as $\hat{z}$ or $-\hat{z}$, and then prove that the general choice (1.63) yields the same magnetic field (1.63) by implementing the rotation of $\hat{z}$ axis to $\mathbf{n}_0$.[7]

In summary, the Berry phase of the spin can be cast as the AB phase factor $L_{\rm B} = \mathbf{A}\cdot\dot{\mathbf{n}}$ of a particle moving on the surface of a unit sphere. The corresponding vector potential is the same as that of a magnetic monopole of charge $\hbar S$, as the relation (1.64) shows.

[7]When not restricted to the surface of the sphere in three-dimensional space, the vector potential $\mathbf{A} = -(\hat{r}_0\times\hat{r})/(r(1-\hat{r}_0\cdot\hat{r}))$ yields a magnetic field of the form $\boldsymbol{\nabla}\times\mathbf{A} = \hat{r}/r^2$.

1.4.4 Action Summary

We have found many different ways to express the same geometric action for spin in this chapter. They are summarized definitively below ($\hbar \equiv 1$):

$$\begin{aligned} S_{\rm B} &= S\int dt(\cos\theta - 1)\partial_t\phi && \text{(angle)} \\ S_{\rm CP^1} &= 2iS\int dt\ \mathbf{z}^\dagger\partial_t\mathbf{z} && (\text{CP}^1) \\ S_{\rm WZ} &= S\int dt\int_0^1 du\ \ \mathbf{n}\cdot(\partial_t\mathbf{n}\times\partial_u\mathbf{n}) && \text{(WZ)} \\ S_{\rm MM} &= \int dt\ \mathbf{A}\cdot\partial_t\mathbf{n} && \text{(monopole)}. \end{aligned} \tag{1.65}$$

Depending on the problem at hand, one way of writing the geometric action may prove more useful than others.

We conclude this chapter with a nice identity that will be used repeatedly in the following chapters. Since all four geometric actions listed above are identical, their variation must also yield identical results. In particular, we saw that varying the Wess–Zumino action yielded (1.59). The same conclusion must also be reached from the variation of $S_{\rm CP^1} = -2S\int dt\ a_t$, i.e., $\delta a_t/\delta\mathbf{n} = (1/2)\partial_t\mathbf{n}\times\mathbf{n}$. This equality arose from purely mathematical considerations, however, and, therefore, must extend to other derivatives as well. In general, we may write

$$\frac{\delta a_\mu}{\delta\mathbf{n}} = \frac{1}{2}\partial_\mu\mathbf{n}\times\mathbf{n}, \tag{1.66}$$

where $a_\mu = -i\mathbf{z}^\dagger\partial_\mu\mathbf{z}$.

References

1. Landau, L.D., Lifshitz, E.M.: Theory of the dispersion of magnetic permeability in ferromagnetic bodies. Phys. Z. Sowietunion **8**, 153 (1935)
2. Berry, M.V.: Quantal phase factors accompanying adiabatic changes. Proc. R. Soc. Lond. A **392**, 45 (1984)
3. Witten, E.: Global aspects of current algebra. Nuc. Phys. B **223**, 422 (1983)
4. Thouless, D.J., Kohmoto, M., Nightingale, M.P., Den Nijs, M.: Quantized Hall conductance in a two-dimensional periodic potential. Phys. Rev. Lett. **49**, 405 (1982)

Chapter 2
Solitons and Homotopy

This chapter attempts to guide the readers through several concepts related to homotopy using explicit examples. Many familiar and some unfamiliar topological objects will be introduced along the way. Kinks in one space dimension and vortex in two dimensions are both characterized by the homotopy of a circle. Skyrmions in two, and hedgehogs in three dimensions are objects tied to the homotopy of a sphere. Tony Skyrme's original vision of treating elementary particles by invoking the topology of a three-sphere only arose in three dimensions. Although magnetic systems in condensed matter have not quite realized Skyrme's vision, ample sightings of two-dimensional skyrmions exist today.

2.1 Kink in One Dimension

Consider a pair of real-valued fields $\mathbf{n} = (n_1, n_2)$ that depends on time t and one spatial coordinate x. A simple Lagrangian that is symmetric under the O(2) rotation $\mathbf{n} \rightarrow \mathscr{R}\mathbf{n}$ can be constructed from the combinations of $(\partial_\mu \mathbf{n}) \cdot (\partial_\mu \mathbf{n})$ and $\mathbf{n} \cdot \mathbf{n}$ terms $(\mu = t, x)$:

$$L = \frac{1}{2} \int dx \left[(\partial_t \mathbf{n})^2 - (\partial_x \mathbf{n})^2 - m^2 \mathbf{n} \cdot \mathbf{n} \right], \tag{2.1}$$

where, for convenience, all the dimensional factors are set equal to unity. The associated equation of motion is a linear partial differential equation,

$$(-\partial_t^2 + \partial_x^2 - m^2)\mathbf{n} = 0, \tag{2.2}$$

which qualifies the Lagrangian (2.1) as an example of a *linear* field theory. The particular example we have here is known as the Klein-Gordon equation.

J.H. Han, *Skyrmions in Condensed Matter*, Springer Tracts in Modern Physics 278, https://doi.org/10.1007/978-3-319-69246-3_2

The action $S = \int dt L$ remains invariant under the translation $t \to t + \delta t$, and $x \to x + \delta x$. Both of these symmetries lead to corresponding conservation laws in accordance with Noether's theorem,

$$\begin{aligned} &\partial_t[(\partial_t \mathbf{n})^2 + (\partial_x \mathbf{n})^2 + m^2\mathbf{n}^2] + \partial_x[-2\partial_t \mathbf{n} \cdot \partial_x \mathbf{n}] = 0 \;\; (t \to t + \delta t), \\ &\partial_t[-2\partial_t \mathbf{n} \cdot \partial_x \mathbf{n}] + \partial_x\left[(\partial_t \mathbf{n})^2 + (\partial_x \mathbf{n})^2 - m^2\mathbf{n}^2\right] = 0 \;\; (x \to x + \delta x), \end{aligned} \tag{2.3}$$

and can be derived by multiplying the equation of motion (2.2) by $\partial_t \mathbf{n}$ and $\partial_x \mathbf{n}$, respectively.

On the other hand, what we would really like to consider in this and the following few sections are *nonlinear* field theories. One can certainly construct a nonlinear field theory by adding an interaction term such as $(\mathbf{n} \cdot \mathbf{n})^2$ to the linear model above. Alternatively, one can instead explore a more drastic route, such as that followed by T. Skyrme, in which the nonlinearity is introduced by a *constraint* on the field:

$$\mathbf{n} \cdot \mathbf{n} = n_1^2 + n_2^2 = 1. \tag{2.4}$$

Now, the magnitude of the field must be unity everywhere in spacetime, which makes $\mathbf{n}$ a point on the unit circle, S^1. Owing to the constraint, the actual degrees of freedom of the field are reduced from two (n_1 and n_2) to just one (i.e., angle θ). One can thus express $\mathbf{n}$ in terms of the angle θ in a way that automatically takes care of the constraint,

$$\mathbf{n} = (\cos\theta, \sin\theta), \tag{2.5}$$

and construct a theory in terms of the "free" dynamical variable θ. In doing so, $(\partial_\mu \mathbf{n}) \cdot (\partial_\mu \mathbf{n})$ reduces to $(\partial_\mu \theta)^2$, while $\mathbf{n} \cdot \mathbf{n}$ becomes a constant. The Lagrangian (2.1) then turns into

$$L = \frac{1}{2} \int dx \, [(\partial_t \theta)^2 - (\partial_x \theta)^2], \tag{2.6}$$

and the two conservation laws derived in (2.3) are accordingly reduced to

$$\begin{aligned} &\partial_t[(\partial_t \theta)^2 + (\partial_x \theta)^2] + \partial_x[-2\partial_t \theta \cdot \partial_x \theta] = 0 \;\; (t \to t + \delta t), \\ &\partial_t[-2\partial_t \theta \cdot \partial_x \theta] + \partial_x\left[(\partial_t \theta)^2 + (\partial_x \theta)^2\right] = 0 \;\; (x \to x + \delta x). \end{aligned} \tag{2.7}$$

Skyrme, in a series of publications starting from the late 50s,[1] noted that another kind of conservation law, quite unrelated to the symmetry principle, existed for the

[1] Skyrme's idea of a topological field theory, topological conservation law, and their application to elementary particles have been developed over half a dozen papers [1–5]. Although Ref. [2] receives the most citation from modern readers, it is worth looking through his earlier publication Ref. [1] that dealt with the one-dimensional example, where he must have developed much of his core intuition. The solution of what is nowadays known as the sine-Gordon model is also found in this early paper.

kinds of nonlinear field theories described by (2.6). This new kind of conserved two-current (i.e., one time, one space) could be constructed as

$$J_\alpha = \frac{1}{2\pi}\varepsilon_{\alpha\beta}\varepsilon_{ab}n_a\partial_\beta n_b. \tag{2.8}$$

Why there is the factor 2π in the denominator will be explained shortly. The first of the two antisymmetric tensors, $\varepsilon_{\alpha\beta}$, refers to the spacetime index (t, x), while the second tensor ε_{ab} refers to the internal indices of the field $\mathbf{n} = (n_1, n_2)$. On account of the manner in which this current was defined, the spacetime in question can only be two-dimensional, i.e., (1+1)-dimensional. The topological current given above thus reduces to one temporal (J_t), and one spatial (J_x) component:

$$J_t = \frac{1}{2\pi}\partial_x\theta, \quad J_x = -\frac{1}{2\pi}\partial_t\theta, \tag{2.9}$$

with the obvious consequence that $\partial_t J_t + \partial_x J_x = 0$. This is the continuity equation for the putative "charge" density J_t and "current" density J_x. Furthermore, the conservation of the topological current can be proven without recourse to the explicit parametrization (2.5) by noting

$$\partial_\mu J_\mu = \frac{1}{2\pi}\varepsilon_{\mu\nu}\varepsilon_{ab}\partial_\mu n_a\partial_\nu n_b = \frac{1}{\pi}\frac{\partial[n_1, n_2]}{\partial[t, x]}. \tag{2.10}$$

That is, the divergence $\partial_\mu J_\mu$ is simply the Jacobian of the mapping from the two-dimensional coordinates (t, x) to (n_1, n_2). One learns in calculus that the Jacobian measures the area of a small covered patch in the target space, in this case (n_1, n_2), as the base space coordinates (t, x) cover an area $\Delta t\Delta x$. However, since $\mathbf{n}$ can only trace out a trajectory on a circle due to the constraint $n_1^2 + n_2^2 = 1$, there is no area element to be covered in the target space and the Jacobian has to be zero.

We would like to get a better understanding of the nature of the conserved charge, so with this in mind, if we write $J_t = \rho = (1/2\pi)\partial_x\theta$ as a density, the total "charge" associated with a particular field configuration $\theta(x, t)$ is

$$Q(t) = \int_{-\infty}^{\infty}\rho(x, t)dx = \frac{1}{2\pi}\int_{-\infty}^{\infty}\partial_x\theta(x, t)dx = \frac{1}{2\pi}[\theta(\infty, t) - \theta(-\infty, t)]. \tag{2.11}$$

On account of the conservation law just derived, this charge has to be a constant of motion, $Q(t) = Q$. Let us imagine that we impose the periodic boundary condition, requiring $\mathbf{n}(\infty, t) = \mathbf{n}(-\infty, t)$. Then, the angular variable $\theta(\infty, t)$ needs to match the value at $x = -\infty$, i.e., $\theta(-\infty, t)$, up to a multiple of 2π:

$$\theta(\infty, t) = \theta(-\infty, t) + 2\pi N, \quad N = \text{integer}. \tag{2.12}$$

Thanks to the division by 2π in the definition (2.8), the charge of the system is nothing other than the integer winding number accumulated across the one-dimensional strip: $Q = N$.

The argument that led to the association of the conserved charge Q with the winding number N was purely mathematical, in the sense that no knowledge of the Hamiltonian or the dynamics of the field was required to reach the conclusion; such an argument is referred to as *topological*. Clearly, the topological conservation law is different from symmetry-originated conservations in that one does not even need to know the Hamiltonian or its internal symmetries. Skyrme's epiphany was to associate the topological number in a nonlinear field theory with the quantum number of sub-atomic particles. The traditional viewpoint of quantum field theory is that an elementary particle is described by a field variable, and the particle's Lagrangian is written out in terms of such a field variable. In Skyrme's interpretation, on the other hand, the field itself is not an elementary particle but a particular configuration of the field with a nonzero topological number is. In essence, Skyrme proposed the radical view that a particle is an "emergent feature" of some underlying field configuration, rather than being defined by the field itself.

In order to qualify a certain object of field theory as a particle, however, one must do more than simply show the existence of a conserved integer charge. A particle is a point-like object after all, and that feature is expected of any "reasonable" model of elementary particles. To help us build our intuition, let us consider a simple field configuration, reasonably localized in space, and carrying a nonzero charge $Q = 1$:

$$\theta(x) = \pi \tanh(x/\lambda) . \tag{2.13}$$

This configuration varies from $-\pi$ at $x = -\infty$ to $+\pi$ at $x = +\infty$, regardless of the value for λ, which represents the width over which there is a significant variation of the angle θ. The width λ also covers the extent of the particle density $\rho(x)$. Although the topological charge Q of the soliton is independent of λ, its energy does depend on λ,

$$H = \frac{1}{2}\int (\partial_x\theta)^2 dx = \frac{\pi^2}{2\lambda^2}\int \left[\coth\left(\frac{x}{\lambda}\right)\right]^2 dx = \frac{\pi^2}{2\lambda}\int (\coth y)^2 dy. \tag{2.14}$$

Since the energy diminishes as the width λ is increased, the most energetically stable soliton configuration is thus the one where $\lambda = \infty$, which is unfortunately nothing like a localized object one expects of a particle! In fact, the argument is even more general than this explicit example suggests, since any function $\theta(x) = f(x/\lambda)$ that is typified by a certain width λ will be subject to the same dimensional relation:

$$H = \frac{1}{2\lambda}\int_{-\infty}^{\infty} (\partial_y f(y))^2 dy \propto \frac{1}{\lambda}. \tag{2.15}$$

This is an instance of what is known as the Derrick–Hobart theorem [6, 7]. In a simple field theory such as $H = (1/2)\int dx(\partial_x\theta)^2$, the soliton does enjoy the topological protection of its charge, but not the kind of energetic stability that would preserve its locality in space.

In order to ensure that a given nontrivial soliton configuration remains energetically stable as well as topologically protected, some additional work needs to be done. To be precise, additional terms are required to prevent the width λ from diverging, so that these additional terms cause the energy to have a nonmonotonic dependence on λ. A magnetic analogy can help decide just which term will get the job done. Taking $\mathbf{n} = (n_1, n_2)$ as a two-dimensional vector defining the magnetic orientation, one can think of adding additional energy terms through an external field $-\mathbf{B} \cdot \mathbf{n}$, higher powers of the gradient $(\partial_\mu \mathbf{n})^4$, or simply higher powers of $\mathbf{n}$. For now we focus on the last possibility, and also impose that the additional terms have symmetries under the time-reversal $\mathbf{n} \to -\mathbf{n}$ and rotation by 90°, $(n_1, n_2) \to (n_2, -n_1)$. The lowest order term satisfying these conditions is $n_1^4 + n_2^4$. With this new term, the Lagrangian reads

$$
\begin{aligned}
L &= \frac{1}{2} \int dx \left[(\partial_t \mathbf{n})^2 - (\partial_x \mathbf{n})^2 - \frac{1}{8} m^2 (n_1^4 + n_2^4) \right] \\
&= \frac{1}{2} \int dx \left[(\partial_t \theta)^2 - (\partial_x \theta)^2 - \frac{1}{8} m^2 (1 - \cos 4\theta) \right].
\end{aligned} \tag{2.16}
$$

Viewing $V(\theta) \propto (1 - \cos 4\theta)$ as a potential energy for θ, there is an obvious pinning effect that tries to fix the orientation angle of $\mathbf{n}$ at multiples of $\pi/2$.

With this interaction, the modified equation of motion is

$$
(\partial_t^2 - \partial_x^2)\theta + \frac{1}{4} m^2 \sin 4\theta = 0. \tag{2.17}
$$

This is the celebrated sine-Gordon equation, one of the best-known examples of a nonlinear field equation with a known exact solution. The "sine" part of the nomenclature comes from the sine term in the equation, while the "Gordon" is derived from the earlier Klein-Gordon equation, which is the linearized version obtained by replacing $\sin 4\theta$ with 4θ.[2] The sine-Gordon equation reduces to the well-known pendulum equation $\partial_t^2 \theta + m^2 \sin 4\theta = 0$, in the uniform limit $\theta(x, t) = \theta(t)$. On the other hand, if we look for a static solution, $\theta(x, t) = \theta(x)$, the equation permits an exact solution (Fig. 2.1)

$$
\begin{aligned}
&-\partial_x^2 \theta + m^2 \sin 4\theta = 0 \\
&\theta(x) = \tan^{-1}[\exp(mx)].
\end{aligned} \tag{2.18}
$$

Depending on the sign of m, $\theta(x)$ evolves from 0 to $\pi/2$ (for $m > 0$) as a quarter-charged soliton, or from $\pi/2$ to 0 as an anti-soliton. The $Q = \pm 1/4$ charge can easily be re-labeled as integers ± 1 by defining 4θ as the new angle θ. Equally popular names for one-dimensional topological solitons and anti-solitons are the "kink" and

[2] To do justice to Skyrme, this equation should really be called Klein-Gordon-Skyrme equation.

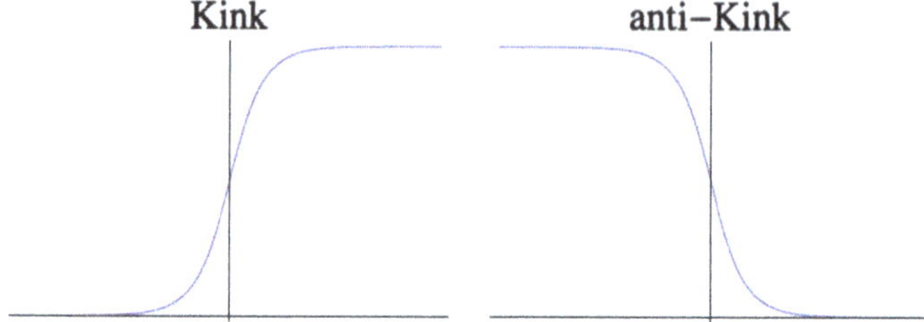

Fig. 2.1 Kink and anti-kink profiles

"anti-kink". The nature of the kink solution is such that it connects one minimum of the potential $V(\theta) \propto -\cos 4\theta$ to another. The full time-dependent solution is obtained by the boost: $x \to \gamma(x \pm vt)$, where v is less than unity, and γ is the Lorentz factor $\gamma = (1 - v^2)^{-1/2}$. The kink solution carries a well-defined topological number Q, a well-defined shape $\theta(x \pm vt)$ that remains constant over all times, and therefore embodies the character of an elementary particle.

We have seen that the topological charge of the kink was obtained by integrating $\partial\theta/\partial x$ over the entire one-dimensional space. This space can be wrapped into a ring by identifying the two ends of a chain of length L as equal. On the other hand, the angle θ also refers to a point on the unit circle, and the whole function $\theta(x)$ can be viewed as a mapping from one circle to another:

$$\theta : S^1 \to S^1. \tag{2.19}$$

The first space with coordinate x of this mapping is called the "base" space, while the second space is called the target space. We note that the winding number embodied by the function $\theta(x)$ expresses the number of times the target space S^1 is covered, as the base space S^1 is traversed only once. By its nature, the answer must be an integer. Mathematicians have expressed such a phenomenon with the notation

$$\pi_1(S^1) = \mathbb{Z}, \tag{2.20}$$

where the subscript 1 in π_1 signifies that the base space is a circle S^1, while the S^1 in the argument signifies that the unit circle S^1 is also the target space of field θ. Finally, $\mathbb{Z}$ signifies that an integer number characterizes the nature of the wrapping.

A few years after Skyrme completed his vision to regard a kink in the one-dimensional field profile as a particle, he was able to successfully generalize the kink $\leftrightarrow$ particle correspondence to three space dimensions in a tour-de-force demonstration of his mathematical prowess. Unlike Skyrme, we mere mortals will instead learn how this was achieved by increasing the space dimension by one dimension at a time, each time only slightly generalizing the strategy garnered from the study of the reduced dimensionality systems.

2.2 Vortices in Two Dimensions

The two-component unit vector field in one-space, one-time dimension supported a topological object called the kink. The topology underlying the quantization of charge was that of mapping S^1 to S^1. In this section, another topological object formally characterized by the same homotopy $\pi_1(S^1)$ will be studied. This time, it is a singular object existing in one higher space dimension than before, and the circle forming the base space corresponds to an ordinary circle $x^2 + y^2 = R^2$ (or a closed path topologically equivalent to such a circle) in two dimensions. We will not be concerned with the time dependence of the field.

In this case, the unit vector field $\mathbf{n}(x, y)$ depends on two-dimensional coordinates (x, y), and if one walks along a circle covering the base space S^1 and records how the $\mathbf{n}$-vector field defined on that circle winds, there comes a realization that the the answer must again be an integer, thanks to the same homotopy consideration $\pi_1(S^1) = \mathbb{Z}$. Suppose a particular base circle yielded a nonzero winding number +1 for the $\mathbf{n}$-vector field defined on it. Then this number, being a topological integer, must remain the same as one shrinks or expands the diameter of the circle, since the $\mathbf{n}$-field can only vary smoothly (by assumption) and a smooth change cannot bring about a change of one integer to another integer.

This kind of field configuration with a nonzero winding number in two dimensions is called a vortex. An explicit example for $N = +1$ is

$$\mathbf{n}_v(x, y) = \left(\frac{x}{r}, \frac{y}{r}\right). \tag{2.21}$$

For any fixed radius $r = \sqrt{x^2 + y^2} = R$ this reduces to $\mathbf{n} = (\cos\varphi, \sin\varphi)$, where the cylindrical angle is defined by $\varphi = \tan^{-1}(y/x)$. In fact this is just an identity map from one unit circle to another, $\mathbf{n}(\hat{r}) = \hat{r}$, where $\hat{r}$ indicates the position on the base circle S^1. In order to create a vortex with winding number N, one can use the field configuration

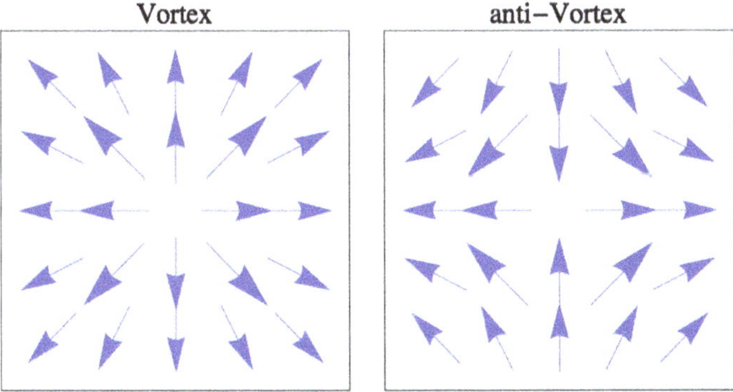

Fig. 2.2 Vortex and anti-vortex configurations in two dimensions

$$\mathbf{n}_v = (\cos[N\varphi], \sin[N\varphi]) \,, \tag{2.22}$$

where N is any positive or negative integer. Configurations with negative values of N are called anti-vortices (Fig. 2.2). We note that any other vortex or anti-vortex configuration with the same topological number N must be able to smoothly deform into the form given above.

In order to compute the topological number for the field configuration (2.22), we adopt the same formula for the topological current as in (1+1)-dimensions, i.e., (2.9). Instead of t and x components, however, we now have x and y components of the current:

$$J_x = \frac{1}{2\pi}\partial_y\theta, \quad J_y = -\frac{1}{2\pi}\partial_x\theta. \tag{2.23}$$

The topological current density associated with a vortex configuration with winding number N follows from $\theta = N\varphi = N\tan^{-1}(y/x)$:

$$\mathbf{J} = (J_x, J_y) = \frac{N}{2\pi}\left(\frac{x}{r^2}, \frac{y}{r^2}\right). \tag{2.24}$$

As we know from vector calculus, the two-dimensional divergence of $\mathbf{J}$ is a delta function, $\nabla \cdot \mathbf{J} = N\delta^2(\mathbf{r})$, and hence, the integral of the divergence, by way of Stokes' theorem, gives

$$\int dxdy \, \nabla \cdot \mathbf{J} = \oint (J_x dy - J_y dx) = \frac{1}{2\pi}\oint d\theta = N. \tag{2.25}$$

This is the same integer winding number as the one obtained from the one-dimensional kink in (2.11).

The singularity in the vortex configuration arises from the fact that the winding has to occur more and more rapidly as the radius of the loop is reduced to zero. No such operation is required for the kink, which winds around a loop of fixed size. The singular nature of the vortex configuration is manifest in the divergent strength of the current density $|\mathbf{J}| \sim 1/r$, and can be seen from an energetic point of view since the natural energy functional for the $\mathbf{n}$-vector field in two space dimensions is

$$H = \frac{1}{2}\int dxdy \, (\nabla\theta)^2 = 2\pi^2 \int dxdy \, \mathbf{J}^2. \tag{2.26}$$

Since $|\mathbf{J}| \sim 1/r$, the energy density of a vortex configuration is proportional to $1/r^2$, which leads to the logarithmic singularity of the integral both with respect to the linear dimension L of the sample being integrated over, and with respect to the the core radius a down to which the integration loop shrinks. $E_v \sim N^2 \log(L/a)$.

One can go on to write down a multi-vortex configuration thanks to the additive property of the angle θ, i.e., $\theta(x, y) = \sum_i \theta(x - X_i, y - Y_i)$. Here, $\mathbf{R}_i = (X_i, Y_i)$ indicates the core of each vortex carrying the winding number N_i, and one may write $\theta_i(x - X_i, y - Y_i) = N_i \tan^{-1}[(y - Y_i)/(x - X_i)]$. The topological continuity equation then becomes

$$\nabla \cdot \mathbf{J} = \sum_i N_i \delta^2(\mathbf{r} - \mathbf{R}_i), \tag{2.27}$$

with a source term appearing now on the right. This is just the two-dimensional version of the Gauss's law in electrodynamics. Indeed, the topological charge N_i plays the same role as the electric charge, strengthening the view that a topological object (in this case a vortex) behaves like an elementary particle. Quantized vortices are familiar objects in quantum condensates such as superfluids, superconductors, and most recently, in quantum gases. The fundamental dynamics of vortex motion will be discussed in Chap. 4.

2.3 Skyrmion in Two Dimensions

2.3.1 *Formal Aspects*

The elements of the two-component unit vector field $\mathbf{n}$ correspond to elements of the U(1) group, which is otherwise known as the SO(2) group. This correspondence to the U(1) group may be seen explicitly since an arbitrary $\mathbf{n}$-vector can adopt an equivalent complex notation, $n_1 + in_2 = e^{i\theta}$, which is an element of the U(1) group. Similarly, the correspondence to the SO(2) group arises because an arbitrary $\mathbf{n}$-vector may be written as the result of the SO(2) rotation matrix,

$$\mathscr{R}(\theta) = \begin{pmatrix} \cos\theta & -\sin\theta \\ \sin\theta & \cos\theta \end{pmatrix},$$

acting on the reference unit vector $\mathbf{n}_0 = \begin{pmatrix} 1 \\ 0 \end{pmatrix}$, thus establishing a one-to-one correspondence between a vector $\mathbf{n}$ and an element of the SO(2) group. The two spaces (i.e., the space of two-component unit vectors and space of SO(2) rotation matrices) may therefore be considered to be equivalent. In other words, the two groups are isomorphic: $\mathrm{U}(1) \simeq \mathrm{SO}(2)$.

After our thorough investigation of topological objects arising from the two component unit vector field (kinks and vortices), it seems natural to consider the topological field theory of a three-component unit vector field $\mathbf{n} = (n_1, n_2, n_3)$. Each field vector $\mathbf{n}$ defines a point on the unit sphere S^2 (which is just another name for the soccer ball). Just as an SO(2) rotation of the reference vector could produce an arbitrary unit vector on the circle, an arbitrary point on the two-sphere is produced by the SO(3) rotation of the reference vector $\mathbf{n}_0 = (0, 0, 1)$. A simple way to see this is to operate on $\mathbf{n}_0$ directly with the general SO(3) matrix

$$\mathscr{R}(\alpha, \beta, \gamma) = e^{-i\alpha S^z} e^{-i\beta S^y} e^{-i\gamma S^z}. \tag{2.28}$$

The first operation $e^{-i\gamma S^z}$ on $\mathbf{n}_0$ produces no effect at all, and the remaining operations yield

$$e^{-i\alpha S^z} e^{-i\beta S^y} \mathbf{n}_0 = \mathbf{n} = (\sin\beta\cos\alpha, \sin\beta\sin\alpha, \cos\beta), \tag{2.29}$$

a point on S^2.

It should be cautioned, however, that the space of unit vectors on S^2 is not the same as the space of SO(3) rotation matrices. While the full set of Euler rotations $\mathscr{R}(\alpha, \beta, \gamma)$ forms an SO(3) group, as we just showed, an element of this group acting on the reference vector $\mathbf{n}_0$ produces the same vector $\mathbf{n}$, an element of S^2, no matter what value is chosen for the angle γ. It is as if different elements of $e^{-i\gamma S^z}$, which are called the "fiber", can be "bundled" together and treated as one element, say an element with $\gamma = 0$. The bundled collection of fibers is known as the "fiber bundle", or the "coset space", and is denoted by G/H, where, in this particular example, G is the SO(3) rotation group and H is its U(1) subgroup. The unit sphere S^2 is thus equivalent to the coset space SO(3)/U(1),[3]

$$S^2 \simeq \mathrm{SO}(3)/\mathrm{U}(1). \tag{2.30}$$

In defining higher-dimensional topological numbers, one needs to enlarge the base space as well as the target space of the vector field. The base space in which kinks were found was one-dimensional; however, the smallest base space in which a stable topological configuration of an $\mathbf{n} \in S^2$ vector can exist is two-dimensional. The two-dimensional Euclidean space of (x, y) coordinates, called $\mathbb{R}^2$ in the math vocabulary, is not the same as the two-sphere. Nevertheless, if the vector field $\mathbf{n}$ assumes the same value, say $\mathbf{n}_0$, as $R = \sqrt{x^2 + y^2} \to \infty$, it becomes possible to treat the entire circumference of the two-dimensional Euclidean plane as effectively one point. The idea is similar to the process of wrapping a sheet of dumpling pastry into a sphere. The dumpling-making process is complete when the perimeters of the original flat sheet are merged together at the apex of a finished dumpling. Instead of using an actual flour sheet, we will show how to transform $\mathbb{R}^2$ into S^2 by the procedure called the stereographic projection. With both the base and target spaces established as that of S^2, the topological number is the number of times the target sphere is wrapped around as one walks around the base sphere. A standard homotopy result $\pi_2(S^2) = \mathbb{Z}$ ensures that the integer winding number characterizes the mapping. In fact, for any S^d to S^d map in d dimensions we have

$$\pi_d(S^d) = \mathbb{Z}. \tag{2.31}$$

[3] A general theorem in topology is that $\mathrm{SO}(n+1)/\mathrm{SO}(n) \simeq S^n$, and $\mathrm{SU}(n+1)/\mathrm{SU}(n) \simeq S^{2n+1}$. The case we are considering here is $n = 2$.

It is a common practice to refer to topologically nontrivial field configurations obtained from sphere-to-sphere mapping in two or three space dimensions as the skyrmion, although Skyrme himself never bothered to worry about the $d = 2$ case.

Another possible characterization of the two-dimensional space may be achieved by imposing periodic boundary conditions in both the x and y directions, thereby turning $\mathbb{R}^2$ into a two-torus $\mathbb{T}^2$. At first sight, the homotopy group for the mapping of two-torus to the two-sphere seems more involved than that of S^2 to S^2. Luckily, there is a theorem stating that all homotopy mappings from the torus $\mathbb{T}^2$ to a target manifold can be classified as two independent π_1 mappings and one π_2 mapping [8]. Since $\pi_1(S^2) = 0$, the only non-trivial homotopy map from the two-torus to the two-sphere is still $\pi_2(S^2) = \mathbb{Z}$, the same result that we obtained from compactifying $\mathbb{R}^2$ directly into S^2.

2.3.2 First Encounter with Skyrmions

With the formal issue of the order parameter space out of the way, we may start asking what kind of interesting topological structures are available for a vector $\mathbf{n}$ residing on S^2. In line with the homotopy result $\pi_2(S^2) = \mathbb{Z}$, we first seek to construct a topological structure in two dimensions explicitly.

Guided by our previous exercise in (1+1) dimensions, we begin by writing a topological current density vector

$$J_\alpha = \frac{1}{8\pi}\varepsilon_{\alpha\beta\gamma}\varepsilon_{abc}n_a\partial_\beta n_b\partial_\gamma n_c. \tag{2.32}$$

We note that the number of spacetime indices α, β, γ and field components a, b, c have both increased by one over the previous (1+1)-dimensional case. The topological continuity equation $\partial_\alpha J_\alpha = 0$ follows from the fact that

$$\partial_\alpha J_\alpha \propto \frac{\partial[n_1, n_2, n_3]}{\partial[t, x, y]}. \tag{2.33}$$

This quantity once again vanishes due to the constraint $\mathbf{n}^2 = 1$, which reduces the number of the field's degrees of freedom by one. As before, the conserved charge is defined as the integral of the J_t component, given by

$$J_t = \frac{1}{8\pi}\varepsilon_{abc}n_a(\partial_x n_b\partial_y n_c - \partial_y n_b\partial_x n_c) = \frac{1}{4\pi}\mathbf{n}\cdot\left(\frac{\partial\mathbf{n}}{\partial x}\times\frac{\partial\mathbf{n}}{\partial y}\right). \tag{2.34}$$

The spatial components of the topological current may be obtained by the cyclic permutation of t, x, y. The two-dimensional integral of this, $Q = \int dxdy J_t(x, y, t)$, is a constant of motion on account of the topological conservation law, $\partial_\alpha J_\alpha = 0$. The integer number Q counts how many times the $\mathbf{n}$-vector covers the target sphere while the (x, y) coordinate space is covered. In the literature, the field configurations with nonzero values of Q are called skyrmions, or sometimes, baby-skyrmions to distinguish them from their three-dimensional cousins.

Expressing the unit vector field with spherical coordinates, $\mathbf{n} = (\sin\beta\cos\alpha, \sin\beta \sin\alpha, \cos\beta)$, yields an expression for the topological charge density which can prove useful:

$$J_t = \frac{1}{4\pi}\sin\beta\left(\frac{\partial\beta}{\partial x}\frac{\partial\alpha}{\partial y} - \frac{\partial\beta}{\partial y}\frac{\partial\alpha}{\partial x}\right) = \frac{1}{4\pi}\frac{\partial[\alpha, \cos\beta]}{\partial[x, y]}. \tag{2.35}$$

The right-hand side of this equation is simply the Jacobian of the coordinate transformation from two-dimensional Euclidean coordinates (x, y) to the spherical coordinates $(\alpha, \cos\beta)$, divided by 4π, the volume of the unit sphere.

In our (1+1)-dimensional example, the periodic boundary condition effectively reduced the one-dimensional real axis $\mathbb{R}^1$ into a circle, S^1. The quantization of the topological charge Q then followed naturally from the homotopy rule $\pi_1(S^1) = \mathbb{Z}$. Pursuant to this guide and the formal discussion of Sect. 2.3.1, we may map the two-dimensional Euclidean space $\mathbb{R}^2$ into a sphere S^2 by imposing the condition

$$\mathbf{n}(r \to \infty, t) = \mathbf{n}_\infty \quad \left(r = \sqrt{x^2 + y^2}\right). \tag{2.36}$$

The explicit construction of the topological configuration then proceeds in two steps. The first step is the one-to-one mapping of the two-dimensional Euclidean space to S^2: $(x, y) \to \hat{r} \in S^2$. Next, one establishes the identity map $\mathbf{n}(\hat{r}) = \hat{r}$, which should generate a topological configuration with unit winding number automatically. Combining the two mappings, one finds the desired configuration with unit winding number

$$\mathbf{n}(x, y) = \hat{r}(x, y). \tag{2.37}$$

The procedure that accomplishes the first task is called the stereographic projection, and is sketched in Fig. 2.3. A sphere with diameter R sits on the Euclidean plane with its tangent point located at its south pole. Each point in the plane has the cylindrical coordinate representation $(x, y) = r(\cos\varphi, \sin\varphi)$, while each point on the sphere has the spherical coordinate representation $R(\sin\beta\cos\alpha, \sin\beta\sin\alpha, \cos\beta)$. Under the stereographic projection, the angle φ of the Euclidean plane is identified with the azimuthal angle α of the sphere, i.e., $\alpha = \varphi$, and $(\cos\varphi, \sin\varphi) = (x/r, y/r)$. Meanwhile, the polar angle β is mapped to the radius r in the plane by $r/R = \cot(\beta/2)$, as shown in Fig. 2.3. Since

$$\cos\frac{\beta}{2} = \frac{r}{\sqrt{r^2 + R^2}}, \quad \sin\frac{\beta}{2} = \frac{R}{\sqrt{r^2 + R^2}}, \tag{2.38}$$

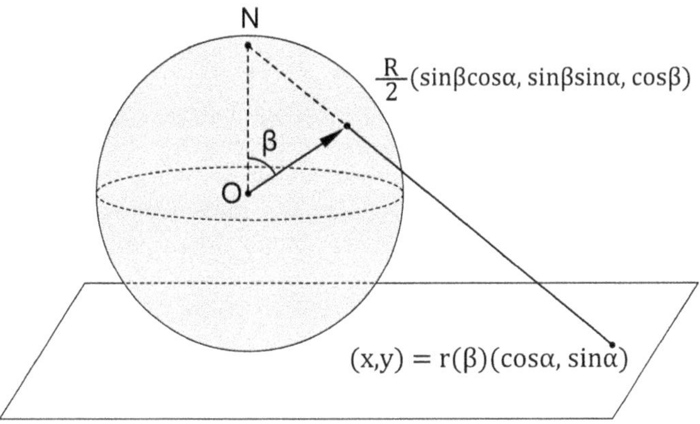

Fig. 2.3 Stereographic projection to the two-dimensional Euclidean plane from a sphere of diameter R. Each point on the sphere is connected in a one-to-one manner to a point on the plane by a ray emanating from the north pole of the sphere. The mapping produces a $Q = -1$ anti-skyrmion

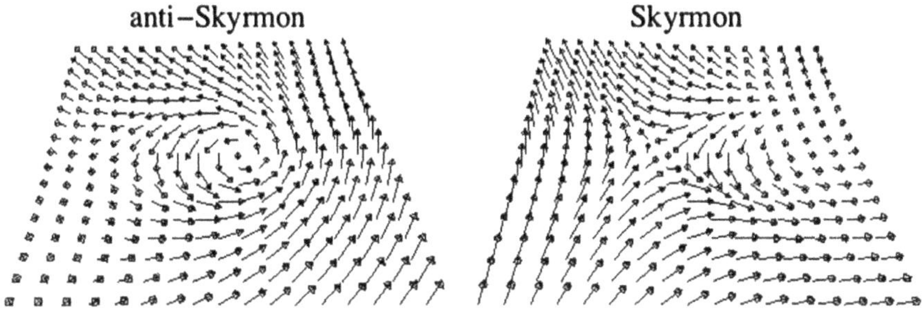

Fig. 2.4 Anti-skyrmion ($Q_s = -1$) and skyrmion ($Q_s = +1$) configurations

we have $\cos\beta = (r^2 - R^2)/(r^2 + R^2)$ and $\sin\beta = 2rR/(r^2 + R^2)$, and the desired mapping is (Fig. 2.4)

$$\mathbf{n}_s(x, y) = \hat{r}(x, y) = \left(\frac{2Rx}{r^2 + R^2}, \frac{2Ry}{r^2 + R^2}, \frac{r^2 - R^2}{r^2 + R^2}\right). \tag{2.39}$$

This is the skyrmion field for a unit winding number, and we can also see that the asymptotic condition (2.36) is satisfied, with $\mathbf{n}_\infty = (0, 0, 1)$. Physical realizations of this kind of spin configuration in magnetic materials are the subject of Chap. 3.

With the construction of skyrmions with unit winding number out of the way, one may ask how to construct skyrmion configurations with an arbitrary integer winding number N. In the previous section on vortices, we had $\mathbf{n}_v = (\cos[N\varphi], \sin[N\varphi])$ for a vortex with winding number N. The corresponding formula for skyrmions can be built upon this expression,

$$\begin{aligned}\mathbf{n}_s &= \big(\sin[f(r)]\mathbf{n}_v, \cos[f(r)]\big)\\ &= \big(\sin[f(r)]\cos[N\varphi], \sin[f(r)]\sin[N\varphi], \cos[f(r)]\big).\end{aligned} \tag{2.40}$$

The skyrmion solution expressed in (2.39) is a special case of this more general case, with $N = 1$ and a particular choice of the radial function $f(r)$, while the vortex configuration corresponds to having $f(r) = \pi/2$ everywhere.

The topological density of the skyrmion configuration (2.40) can be readily computed:

$$\rho_s(x, y) = \frac{1}{4\pi}\mathbf{n}_s \cdot (\partial_x \mathbf{n}_s \times \partial_y \mathbf{n}_s) = \frac{N}{4\pi r} f'(r)\sin[f(r)]. \tag{2.41}$$

Integrated over all space, the topological charge becomes

$$Q_s = \frac{N}{2}\int_0^\infty dr f'(r)\sin[f(r)] = \frac{N}{2}\int_0^\infty dr \frac{dn^z}{dr} = N \cdot \frac{n^z(0) - n^z(\infty)}{2}. \tag{2.42}$$

Depending on whether the vector field varies from north-to-south or south-to-north as r is increased from the origin to infinity, the skyrmion charge Q_s equals the vortex's winding number N up to a sign, i.e., $Q_s = \pm N$. In magnetic literature, the two integer quantities N and $P = [n^z(0) - n^z(\infty)]/2$ are called the vorticity and the polarity, respectively. The skyrmion charge Q_s can be viewed as the product of the two integers

$$Q_s = N \cdot P. \tag{2.43}$$

As the formula suggests, skyrmions of higher topological number can be produced by employing structures with larger vorticities N, or higher polarities P. For instance, the skyrmion state obtained from the stereographic mapping has $Q_s = -1$, due to the fact that $N = 1$ but $P = -1$. Although the expression (2.39) is often said to represent a skyrmion, it is really an anti-skyrmion according to our definition of topological charge.

There is a lesson to be learned from the way the skyrmion solution was written using the vortex solution as a "seed" in (2.40). Recall that the planar character of the spin was the cause of the singularity in our earlier discussion of the vortex solution. The skyrmion configuration lifts this singularity, using the larger space afforded to the vector **n**, by tilting it out of the plane. As a result, the energy of the skyrmion remains finite. We will shortly use this trick once again in order to generate a smooth topological configuration in three dimensions. This will be none other than Skyrme's original vision of the skyrmion.

It was possible to derive the kink solution in (1+1)-dimensions as the saddle-point solution of a specific Hamiltonian. How about the two-dimensional skyrmion? Is there a (classical) Hamiltonian whose saddle-point solution describes the skyrmion? A first guess is, as in the one-dimensional problem,

$$H = \frac{1}{2} \int dxdy \left[(\partial_x \mathbf{n})^2 + (\partial_y \mathbf{n})^2 \right]. \tag{2.44}$$

This kind of model is called the non-linear σ-model, and at first glance, looks like a free quadratic theory. However, this is really not the case due to the $\mathbf{n}^2 = 1$ constraint. Furthermore, it is a scale-invariant theory, in the sense that the coordinate change $(x, y) \to a(x, y)$ leaves the Hamiltonian invariant.

Feeding the skyrmion configuration (2.40) into the Hamiltonian (2.44) yields the skyrmion energy

$$E_s = \pi \int_0^\infty rdr \left[(f')^2 + \frac{N^2}{r^2} (\sin f)^2 \right]. \tag{2.45}$$

The saddle-point equation following from this energy functional,

$$2(rf')' + \frac{N^2}{r^2} \sin 2f = 0, \tag{2.46}$$

determines the shape of the profile function $f(r)$ subject to the boundary conditions at $r = 0$ and $r = \infty$. In particular, the skyrmion solution (2.39), previously obtained from the stereographic projection, is a valid solution of this equation with $(\sin f, \cos f) = (2rR, r^2 - R^2)/(r^2 + R^2)$, regardless of the choice of R. The energy of the skyrmion remains insensitive to R, in keeping with the Derrick-Hobart theorem.

We encountered a similar problem in the construction of the stable kink solution in one dimension. As we saw earlier, the way to beat it was to introduce an additional interaction that could give rise to a nonmonotonic dependence of the energy on the scale parameter R of the topological soliton. In the same manner, we ask what kind of additional terms in the Hamiltonian would fix the radius R? One can begin with the Zeeman interaction that adds the following energy to the skyrmion (assuming $\mathbf{B} = B\hat{z}$)

$$\begin{aligned} E_Z = -\mathbf{B} \cdot \int \mathbf{n} \, dxdy &= -2\pi B \int_0^\infty r \cos \left[f \left(\frac{r}{R} \right) \right] dr \\ &= -2\pi B R^2 \int_0^\infty x \cos[f(x)] dx. \end{aligned} \tag{2.47}$$

This is a quadratic function of R with the sign of the curvature depending on the sign of the integral, and will not in any case yield a stable skyrmion radius. We may try another energy form, this time the quartic term, $\int (n_1^4 + n_2^4 + n_3^4) \, dxdy$; however, dimensional analysis reveals that this too shows a quadratic dependence on R, and so even the combined Zeeman and quartic energy contributions will not be able to fix the skyrmion radius uniquely. By now it should be clear that besides the Zeeman and/or higher-order anisotropy energies, one also needs an energy that depends linearly on R. In Chap. 3 we will show that Dzyaloshinskii-Moriya (DM) interaction in chiral magnets is a critical new interaction that adds just such a linear dependence to the energy.

2.4 Hedgehogs in Three Dimensions

Earlier we discussed how the smooth topological texture in one dimension (kink) and the singular topological texture in two dimensions (vortex) are both described by the same homotopy map $\pi_1(S^1)$. In this section, the analogy is elevated to one higher dimension in an effort to construct a singular topological object in three dimensions that shares the homotopy $\pi_2(S^2)$ with the two-dimensional skyrmion.

Take a sphere $x^2 + y^2 + z^2 = R^2$ embedded in a three-dimensional space and follow the trajectory of the S^2 vector field $\mathbf{n}$ as one walks about the surface of such sphere. A configuration with non-zero integer winding number must exist due to the homotopy relation $\pi_2(S^2) = \mathbb{Z}$. An obvious example is the identity map

$$\mathbf{n}_h = \left(\frac{x}{r}, \frac{y}{r}, \frac{z}{r}\right), \quad r = \sqrt{x^2 + y^z + z^2}. \tag{2.48}$$

This object goes by the amicable name of a "hedgehog", on account of its similarity to the pet of the same name in an irate mood. The topological current density (2.32) of the hedgehog works out to be (replace t with z)

$$\mathbf{J} = (J_x, J_y, J_z) = \frac{1}{4\pi}\frac{\mathbf{r}}{r^3}. \tag{2.49}$$

Viewing $\mathbf{J}$ as a magnetic field evokes the image of a magnetic monopole sitting at the origin. For this reason, some authors prefer to call this configuration a "monopole" rather than a hedgehog.

The energy of the hedgehog (assuming the Hamiltonian to be the nonlinear σ-model),

$$E = \frac{1}{2}\int d^3\mathbf{r}\,[(\partial_x\mathbf{n})^2 + (\partial_y\mathbf{n})^2 + (\partial_z\mathbf{n})^2] = 4\pi\int r^2 dr\left(\frac{1}{r^2}\right), \tag{2.50}$$

diverges as the linear size of the integral $E \sim L$. Besides, the topological current density of the hedgehog satisfies the divergence condition

$$\nabla\cdot\mathbf{J} = \delta^3(\mathbf{r}), \tag{2.51}$$

which suggests its interpretation as a point particle. The surface integral of $\mathbf{J}$, or equivalently the volume integral of $\nabla\cdot\mathbf{J}$, serves as the conserved charge of the hedgehog. In many ways, the hedgehog is the three-dimensional version of the vortex, and an anti-hedgehog state is obtained by simply reversing the arrows, $\mathbf{n}_h \to -\mathbf{n}_h$.

Owing to their singular nature, hedgehogs must co-exist with an equal number of anti-hedgehogs.

Together with the skyrmion of the previous section, we have constructed one nonsingular and one singular configuration of the three-component vector field $\mathbf{n}$, both governed by the same homotopy $\pi_2(S^2)$, in two- and three-dimensional spaces, respectively. Later, when we try to construct a smooth topological configuration in three-dimensional space, the singular field $\mathbf{n}_h$ will be weighted by a function $\sin[f(r)]$ that vanishes at the origin and at infinity, while a fourth component $n_4 = \cos[f(r)]$ will be included to form a four-component unit vector

$$\mathbf{n}_s = \big(\sin[f(r)]\mathbf{n}_h, \cos[f(r)]\big). \tag{2.52}$$

As in the construction of the two-dimensional skyrmion out of the two-dimensional vortex, the idea here is to adopt the singular configuration from the lower $\mathbf{n}$-vector space ($\mathbf{n} \in S^2$) to construct a smooth topological configuration in the higher dimensional space ($\mathbf{n} \in S^3$). Chapter 3 includes a discussion of the hedgehog spin lattice in a three-dimensional model of chiral magnet.

2.5 Skyrmions in Three Dimensions

Armed with these exercises in lesser dimensions, it is time to construct a topological object in three spatial dimensions out of a four-component field $\mathbf{n} = (n_1, n_2, n_3, n_4)$ subject to the unit modulus constraint $\mathbf{n} \cdot \mathbf{n} = 1$. The topological current,

$$J_\alpha = \frac{1}{12\pi^2}\varepsilon_{\alpha\beta\gamma\delta}\varepsilon_{abcd}n_a\partial_\beta n_b\partial_\gamma n_c\partial_\delta n_d, \tag{2.53}$$

is an obvious generalization of the previous topological currents, (2.8) and (2.32).

It is not easy to find examples of a physical system described by a four-component unit vector field in classical physics. However, as noted by Skyrme in his papers, in quantum physics the four numbers $\mathbf{n} = (n_1, n_2, n_3, n_4)$ subject to the unit modulus constraint are equivalent to an element of the SU(2) group. One can verify this statement by arranging the four elements of $\mathbf{n}$ in matrix form as

$$U(\mathbf{n}) = i(n_1\sigma_1 + n_2\sigma_2 + n_3\sigma_3) + n_4 = \begin{pmatrix} n_4 + in_3 & in_1 + n_2 \\ in_1 - n_2 & n_4 - in_3 \end{pmatrix}. \tag{2.54}$$

Here, the three matrices $\boldsymbol{\sigma} = (\sigma_1, \sigma_2, \sigma_3)$ are the familiar Pauli matrices. In order to see that this is an element of the SU(2) group in a transparent fashion, we parameterize the four components of the $\mathbf{n}$ vector as

$$\mathbf{n} = (\mathbf{m}\sin\gamma, \cos\gamma), \tag{2.55}$$

where $\mathbf{m}$ is the unit vector on S^2. The matrix $U(\mathbf{n})$ then assumes the familiar form

$$U = \cos\gamma + i\sin\gamma[\mathbf{m}\cdot\boldsymbol{\sigma}] = \exp\Big(i\gamma[\mathbf{m}\cdot\boldsymbol{\sigma}]\Big), \tag{2.56}$$

of an element of the SU(2) group. Since SU(2) forms a group, so do the points on S^3.

There is an equally good way to express the topological current (2.53) in terms of the SU(2) group element,

$$J_\alpha = -\frac{1}{24\pi^2}\varepsilon_{\alpha\beta\gamma\delta}\mathrm{Tr}\Big[(U^\dagger\partial_\beta U)(U^\dagger\partial_\gamma U)(U^\dagger\partial_\delta U)\Big], \tag{2.57}$$

which one often finds in the field theory literature. The angular representation of the unit vector

$$\begin{aligned}\mathbf{n} &= (\mathbf{m}\sin\gamma, \cos\gamma)\\ &= (\sin\gamma\sin\beta\cos\alpha, \sin\gamma\sin\beta\sin\alpha, \sin\gamma\cos\beta, \cos\gamma)\end{aligned} \tag{2.58}$$

yields yet another equivalent expression for the topological current

$$J_D = \frac{1}{2\pi^2}\varepsilon_{ABCD}\sin\beta(\sin\gamma)^2(\partial_A\alpha)(\partial_B\beta)(\partial_C\gamma), \tag{2.59}$$

where we have introduced capital Roman letters to avoid a notational overlap with the symbol for the angles.

The topological charge follows from integrating the temporal component of the topological current J_t over the three-dimensional space. Following the strategy suggested from the earlier treatments, we may express a three-dimensional skyrmion by

$$\mathbf{n}_s = \Big(\sin[f(r)]\frac{x}{r},\ \sin[f(r)]\frac{y}{r},\ \sin[f(r)]\frac{z}{r},\ \cos[f(r)]\Big). \tag{2.60}$$

Employing the radial dependence given by

$$(\sin f(r), \cos f(r)) = \left(\frac{2rR}{r^2+R^2}, \frac{r^2-R^2}{r^2+R^2}\right), \quad r = \sqrt{x^2+y^2+z^2}, \tag{2.61}$$

in connection to the two-dimensional stereographic projection, actually yields the three-dimensional projection of $\mathbb{R}^3$ to the three-sphere

$$(x,y,z)\in\mathbb{R}^3 \to \left(\frac{2xR}{r^2+R^2}, \frac{2yR}{r^2+R^2}, \frac{2zR}{r^2+R^2}, \frac{r^2-R^2}{r^2+R^2}\right)\in S^3. \tag{2.62}$$

Having constructed a viable skyrmion configuration (2.60), one can calculate its topological charge density,

$$J_t(r) = -\frac{1}{2\pi^2}\frac{(\sin f)^2 f'}{r^2}, \tag{2.63}$$

and the associated topological charge

$$\begin{aligned} Q &= \int_0^\infty J_t(r) 4\pi r^2 dr = \frac{2}{\pi}\int_{f(\infty)}^{f(0)} (\sin f)^2 df \\ &= \frac{1}{\pi}\int_{f(\infty)}^{f(0)} (1-\cos 2f) df \\ &= \frac{f(0)-f(\infty)}{\pi} - \frac{\sin[2f(0)]-\sin[2f(\infty)]}{2\pi}. \end{aligned} \tag{2.64}$$

Let us think about the appropriate boundary conditions for the radial function $f(r)$. The purpose of introducing $f(r)$ was to ensure that the singularity of the hedgehog solution at the origin become invisible, by taking $\sin[f(0)] = 0$. The other property expected of $f(r)$ is to ensure that there is no variation in the **n**-field at spatial infinity to suppress diverging energy, and to ensure this condition, we must take $\sin[f(\infty)] = 0$. In other words, both $f(0)$ and $f(\infty)$ must be some multiples of π, and as a result, the second term in the last line of (2.64) must vanish, ensuring that

$$Q = \frac{f(0)-f(\infty)}{\pi} \tag{2.65}$$

is an integer.

We further note that the skyrmion configuration (2.60) may be expressed in an equivalent SU(2) form by

$$U_s = \exp\left(if(r)\hat{r}\cdot\boldsymbol{\sigma}\right) = \cos[f(r)] + i\hat{r}\cdot\boldsymbol{\sigma}\sin[f(r)]. \tag{2.66}$$

As mentioned earlier, in condensed matter physics, neither the 2×2 unitary matrix nor the unit vector on S^3 is an easily observable quantity. In magnetic models, physical spins have three components at the most. There is, however, a remarkable way to turn an SU(2) field configuration into a corresponding configuration of the vector field on S^2 through a process called the Hopf fibration in mathematics. We will discuss this procedure in the next section, after arming ourselves with some more language related to the CP^1 mapping.

2.6 CP^1 Theory

Let us begin with some mathematical preliminary. A collection of $2n$ real numbers, x_1 through x_{2n}, can be reorganized as a collection of n complex numbers by pairing the number as follows, $z_1 = x_1 + ix_2$, $z_2 = x_3 + ix_4$, etc. If the initial set of real numbers were subject to the unit modular constraint, $\sum_{i=1}^{2n} x_i^2 = 1$, the space of such numbers defines the hypersphere S^{2n-1}. The simplest instance of this occurs for $n = 1$, in which case a point on S^1 is identifiable with a complex number z of unit modulus. For $n = 2$, the four real numbers pair up to give

$$\mathbf{z} = \begin{pmatrix} z_1 \\ z_2 \end{pmatrix} = \begin{pmatrix} x_1 + ix_2 \\ x_3 + ix_4 \end{pmatrix}. \tag{2.67}$$

The normalization condition $\mathbf{z}^\dagger \mathbf{z} = 1$ suggests the possibility of interpreting $\mathbf{z}$ as the wavefunction of a spin-1/2 particle, in which case one also knows that the overall phase does not have a physical impact. This means that a second $\mathbf{z}'$, related to $\mathbf{z}$ by an overall phase factor $\mathbf{z}' = e^{if}\mathbf{z}$, may be viewed as the "same state" as $\mathbf{z}$. This is akin to identifying all the points on the original hypersphere S^3, connected by some $U(1) \simeq S^1$ operation, as one element. This bundling process gives rise to the coset space S^3/S^1, which is the space of allowed wave functions of the two-component spinor. Another name for this space is the "Complex Projective space", or CP space for short. Obviously, the process of pairing up $2n$ real numbers into n complex numbers (or an n-component wave function), and considering the overall phase to be irrelevant, can continue for arbitrary n. The spaces constructed in such a manner, called CP^{n-1}, are given by

$$\mathrm{CP}^{n-1} \simeq S^{2n-1}/S^1, \tag{2.68}$$

in accordance with the argument presented above.

In the previous chapter, the explicit coordinate representation of the CP^1 field

$$\mathbf{z} = \begin{pmatrix} \cos[\beta/2] \\ e^{i\alpha}\sin[\beta/2] \end{pmatrix} \tag{2.69}$$

was used in order to express the geometric phase of the spin action. There, we also saw that the substitution $\mathbf{n} = \mathbf{z}^\dagger \boldsymbol{\sigma} \mathbf{z}$ mapped the spin Hamiltonian into one in terms of $\mathbf{z}$. In particular, the nonlinear σ-model $(\partial_\mu \mathbf{n}) \cdot (\partial_\mu \mathbf{n})$ has an appealing form when written in terms of the CP^1 field:

$$\begin{aligned}(\partial_\mu \mathbf{n}) \cdot (\partial_\mu \mathbf{n}) &= (\partial_\mu \beta)^2 + \sin^2\beta(\partial_\mu \alpha)^2 \\ &= 4(\partial_\mu \mathbf{z}^\dagger)(\partial_\mu \mathbf{z}) - 4([\partial_\mu \mathbf{z}^\dagger]\mathbf{z})(\mathbf{z}^\dagger \partial_\mu \mathbf{z}).\end{aligned} \tag{2.70}$$

Utilizing our previous definition of the emergent gauge field $a_\mu = -i\mathbf{z}^\dagger[\partial_\mu \mathbf{z}] = +i(\partial_\mu \mathbf{z}^\dagger)\mathbf{z}$, it is straightforward to show that the nonlinear σ-model is equivalent to another form,

$$(\partial_\mu \mathbf{n}) \cdot (\partial_\mu \mathbf{n}) = 4\Big([\partial_\mu + ia_\mu]\mathbf{z}^\dagger\Big)\Big([\partial_\mu - ia_\mu]\mathbf{z}\Big) = 4(D_\mu \mathbf{z})^\dagger (D_\mu \mathbf{z}). \quad (2.71)$$

A distinctive feature in rewriting the nonlinear σ-model in terms of the CP^1 field is the appearance of a covariant derivative $D_\mu = \partial_\mu - ia_\mu$, with an abelian gauge field a_μ.

What about the topological current? Is it possible to re-write the (2+1)-dimensional topological current J_α in Eq. (2.32) as a function of the CP^1 field? Given the identity (1.50) previously listed in Chap. 1, the answer is yes, and the current is given by

$$J_\alpha = \frac{1}{4\pi}\varepsilon_{\alpha\beta\gamma}\mathbf{n}\cdot(\partial_\beta \mathbf{n} \times \partial_\gamma \mathbf{n}) = \frac{1}{2\pi}\varepsilon_{\alpha\beta\gamma}\partial_\beta a_\gamma. \quad (2.72)$$

Writing out each component explicitly, the (2+1)-dimensional topological current, expressed in terms of CP^1, is given by

$$\mathbf{J} = \frac{1}{2\pi}\Big(\partial_x a_y - \partial_y a_x, \partial_y a_t - \partial_t a_y, \partial_t a_x - \partial_x a_t\Big) \equiv \frac{1}{2\pi}(b_z, -e_y, e_x). \quad (2.73)$$

In the last expression, we have identified the temporal component of the spacetime current with the "magnetic field" b_z, and identified the two spatial components with the two "electric field" components $(-e_y, e_x)$. The assignment is clearly in keeping with their respective definitions in ordinary electrodynamics. The topological current proves to be none other than the emergent electric and magnetic field in (2+1)-dimensions! The conservation law of the topological current $\partial_\alpha J_\alpha = 0$ then consequently becomes the familiar Faraday's law:

$$\partial_t b_z - \partial_x e_y + \partial_y e_x = 0. \quad (2.74)$$

Thanks to the CP^1 mapping, the theory of topological currents finds an interesting analogy in (2+1)-dimensional electrodynamics.

The two-dimensional skyrmion configuration of charge $Q = N$ found earlier in terms of $\mathbf{n}_s$ (2.40) has a corresponding CP^1 solution

$$\mathbf{z}_s = \begin{pmatrix} \cos[f(r)/2] \\ e^{iN\phi}\sin[f(r)/2] \end{pmatrix}.$$

We will make extensive use of this CP^1 wavefunction in later chapters when trying to discuss the emergent fields associated with both stationary and moving skyrmions.

In (2.72), we showed how to express the topological three-current in terms of the CP^1 field. In fact, it seemed that all the quantities originally expressed in terms of the O(3) vector $\mathbf{n}$ could find an elegant alternative expression in terms of the CP^1 field. We now ask an analogous question for the O(4) vector $\mathbf{n} \in S^3$. How many of the $\mathbf{n}$-labeled quantities will find an elegant expression in terms of the CP^1 field? Given that a SU(2) matrix may be viewed as a rotation matrix in spinor space, we let an arbitrary SU(2) element, e.g., (2.56), act on the reference spinor $\mathbf{z}_0 = \begin{pmatrix} 1 \\ 0 \end{pmatrix}$, to get

$$\mathbf{z} = U\mathbf{z}_0 = \begin{pmatrix} \cos\gamma + i\sin\gamma\cos\beta \\ ie^{i\alpha}\sin\beta\sin\gamma \end{pmatrix}. \tag{2.75}$$

We see that all three angular variables (α, β, γ) remain present in $\mathbf{z}$, and so this remains faithful representation of the SU(2) matrix. This is the O(4) analogue of (2.69), which expressed a given O(3) vector as a CP^1 field. Given this CP^1 representation of SU(2), we may invoke the Hopf map $\mathbf{n} = \mathbf{z}^\dagger \boldsymbol{\sigma}\mathbf{z}$ to calculate the corresponding three-component vector field

$$\begin{aligned} \mathbf{n} &= \left(2[n_1n_3 - n_2n_4], 2[n_1n_4 + n_2n_3], n_3^2 + n_4^2 - n_1^2 - n_2^2\right) \\ &= \Big(2(m_xm_z\sin\gamma - m_y\cos\gamma)\sin\gamma, \\ &\quad 2(m_x\cos\gamma + m_ym_z\sin\gamma)\sin\gamma, \\ &\quad \cos^2\gamma + (m_z^2 - m_x^2 - m_y^2)\sin^2\gamma\Big). \end{aligned} \tag{2.76}$$

Although this is only a three-component vector, all three angles featured in the SU(2) matrix still feature in what is (hopefully) a faithful representation of the SU(2) element.

Continuing with our analysis, we now ask if the topological four-current (2.53) can find an elegant expression using the CP^1 field (2.75). It turns out that by using the gauge field $\mathbf{a} = -i\mathbf{z}^\dagger \nabla \mathbf{z}$ and its curl, one can prove the interesting identity

$$\mathbf{a}\cdot(\nabla\times\mathbf{a}) = -2\sin\beta(\sin\gamma)^2\nabla\alpha\cdot(\nabla\beta\times\nabla\gamma) = -4\pi^2 J_t. \tag{2.77}$$

The result implies that, in terms of the CP^1 gauge field $\mathbf{a}$, the topological charge in three dimensions derived from the four-component unit modular field becomes the simple-looking integral

$$Q = -\frac{1}{4\pi^2}\int dxdydz\, \mathbf{a}\cdot(\nabla\times\mathbf{a}). \tag{2.78}$$

Let us pause for a minute to mull over the meaning of the formula we have just derived. It was mentioned that the three-dimensional skyrmion is hard to realize in condensed matter systems because its construction requires a four-component unit vector field, or a SU(2) matrix field, neither of which are readily available physical fields in condensed matter systems. However, we know that any given unit vector field on S^2 can be decomposed as a CP^1 field by reverse-engineering the Hopf map $\mathbf{n} = \mathbf{z}^\dagger\boldsymbol{\sigma}\mathbf{z}$. In turn, this CP^1 field engenders a gauge field $\mathbf{a} = -i\mathbf{z}^\dagger\nabla\mathbf{z}$. If the gauge field configuration was such that the integral on the right-hand side of (2.78) yielded an integer answer, then by inverting the logic we could claim that the original spin configuration of $\mathbf{n} \in S^2$ was an "incarnation" of the three-dimensional skyrmion! This is possible thanks to the equivalence of the topological number written in terms of the four-component field (2.53), to the one written in terms of what essentially amounts to the three-component field (2.78). This phenomenon is known as the *Hopf fibration*.

As an example, we can apply the observation to the actual skyrmion structure defined earlier. Acting on the reference CP¹ field $\mathbf{z}_0 = \begin{pmatrix} 1 \\ 0 \end{pmatrix}$ with U_s [see (2.66)] yields the CP¹ field for the three-dimensional skyrmion

$$\mathbf{z}_s = U_s \mathbf{z}_0 = \begin{pmatrix} \cos[f(r)] + i \sin[f(r)] z/r \\ i \sin[f(r)](x + iy)/r \end{pmatrix}. \tag{2.79}$$

The classical spin configuration $\mathbf{n}_s$ derived from this CP¹ configuration is given by,

$$\begin{aligned} \mathbf{n}_s &= \mathbf{z}_s^\dagger \boldsymbol{\sigma} \mathbf{z}_s \\ &= \Bigg(2\left[\frac{xz}{r^2} \sin[f(r)] - \frac{y}{r} \cos[f(r)]\right] \sin[f(r)], \\ &\qquad 2\left[\frac{x}{r} \cos[f(r)] + \frac{yz}{r^2} \sin[f(r)]\right] \sin[f(r)], \\ &\qquad \cos^2[f(r)] + \frac{z^2 - x^2 - y^2}{r^2} \sin^2[f(r)] \Bigg). \end{aligned} \tag{2.80}$$

One can also show

$$\mathbf{a}_s \cdot \nabla \times \mathbf{a}_s = \frac{2 \sin^2[f(r)] f'(r)}{r^2}. \tag{2.81}$$

Note how this topological density matches the earlier one (2.63), derived by way of the S³ vector field.

In mathematics, the integer representing the skyrmion number (2.78) is known as the Hopf index. The integral formula (2.78) was discovered by Whitehead [9] as a way to compute the Hopf index explicitly for a given field configuration. Experimentally, observation of the spin structure $\mathbf{n}_s$ (2.80) in a three-dimensional magnet would constitute the discovery of three-dimensional skyrmion structures in condensed matter systems. Such a spin structure is also known as Shankar's monopole [10], and refers to a topological structure realizing the non-trivial $\pi_3(\mathrm{SO}(3)) = \mathbb{Z}$ homotopy class. (Fig. 2.5)

This remark on the homotopy class of Shankar's monopole requires a little more explanation. The SU(2) unitary rotation given by (2.66), acting on the reference CP¹ field $\mathbf{z}_0 = \begin{pmatrix} 1 \\ 0 \end{pmatrix}$, can be thought of as the SO(3) rotation

$$\mathscr{R}_s = \exp\left(2if(r)\mathbf{J} \cdot \hat{r}\right) \tag{2.82}$$

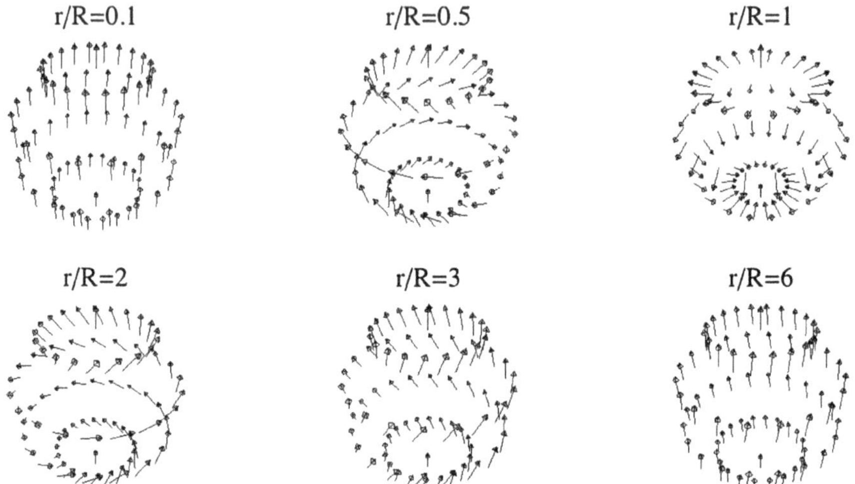

Fig. 2.5 Explicit realizations of Shankar's monopole (2.80) with $f(r) = 2\pi R^2/(r^2 + R^2)$. The spin orientations for a number of fixed values of r/R are shown

acting on the reference spin $\mathbf{n}_0 = (0, 0, 1)$. The result of this operation has already been expressed in (2.80), and we used the same boundary conditions as defined earlier, i.e., both $f(0)$ and $f(\infty)$ are multiples of π. These conditions for the $\mathbf{n}$ vector field imply that $\mathbf{n}(0) = \mathbf{n}(\infty) = \mathbf{n}_0$ for the $\mathbf{n}$ vector field. Thus, the vector field $\mathbf{n}_s = \mathscr{R}_s\mathbf{n}_0$ is an explicit realization of the nontrivial $\pi_3(\mathrm{SO}(3))$ homotopy. The emergent magnetic field $\mathbf{b}_s = \nabla \times \mathbf{a}_s$ associated with the Shankar monopole is also very intricate,

$$\begin{aligned}\mathbf{b}_s = \frac{1}{r^5}\Big(&2\sin f\big[xzr\sin f - r^2(xz\cos f + yr\sin f)f'\big],\\ &2\sin f\big[yzr\sin f - r^2(yz\cos f - xr\sin f)f'\big],\\ &2z^2r(\sin f)^2 + (r^2 - z^2)r^2 f'\sin 2f\Big).\end{aligned} \tag{2.83}$$

Yet another name, the *hopfion*, is associated with the same topological structure in the literature since it realizes a nonzero Hopf index.

One key issue remains to be considered before concluding this chapter–that of finding a model Hamiltonian which supports a hopfion configuration as a metastable state. One such model Hamiltonian, suggested by Faddeev and Niemi [11], consists of simply adding the field energy term to the nonlinear σ-model.

$$H = \frac{1}{2}\int d^3\mathbf{r}\,\sum_{\mu=1}^{3}(\partial_\mu\mathbf{n})^2 + g\int d^3\mathbf{r}\,(\nabla\times\mathbf{a})^2. \tag{2.84}$$

Such a Hamiltonian is not so unrealistic; after all the curl $\nabla \times \mathbf{a}$ is related to the spin texture through

$$\frac{1}{2}\mathbf{n} \cdot (\partial_\mu \mathbf{n} \times \partial_\nu \mathbf{n}) = \varepsilon_{\mu\nu\lambda}(\nabla \times \mathbf{a})_\lambda. \tag{2.85}$$

Finally, we note that a recent breakthrough in creating the hopfion structure in a chiral ferromagnetic fluid and chiral liquid crystal have been reported in Refs. [12, 13]. No realization of a hopfion structure in a hard chiral ferromagnet has been reported to date.

References

1. Skyrme, T.H.R.: A non-linear theory of strong interactions. Proc. R. Soc. Lond. Ser. A. Math. Phys. Sci. **247**, 260 (1958)
2. Skyrme, T.H.R.: A non-linear field theory. Proc. R. Soc. Lond. Ser. A. Math. Phys. Sci. **260**, 127 (1961)
3. Skyrme, T.H.R.: Particle states of a quantized meson field. Proc. R. Soc. Lond. Ser. A. Math. Phys. Sci. **262**, 237 (1961)
4. Perring, J.K., Skyrme, T.H.R.: A model unified field equation. Nuc. Phys. **31**, 550 (1962)
5. Skyrme, T.H.R.: A unified field theory of mesons and baryons. Nuc. Phys. **31**, 556 (1962)
6. Derrick, G.H.: Comments on nonlinear wave equations as models for elementary particles. J. Math. Phys. **5**, 1252 (1964)
7. Hobart, R.H.: On instability of a class of unitary field models. Proc. Phys. Soc. Lond. **82**, 201 (1963)
8. Avron, J.E., Seiler, R., Simon, B.: Homotopy and quantization in condensed matter physics. Phys. Rev. Lett. **51**, 51 (1983)
9. Whitehead, J.H.C.: An expression of Hopf invariant as an integral. Proc. Nat. Acad. Sci. U.S.A. **33**, 117 (1947)
10. Shankar, R.: Application of topology to the study of ordered systems. Le Journal de Physique **38**, 1405 (1977)
11. Faddeev, L., Niemi, A.J.: Stable knot-like structures in classical field theory. Nature **387**, 58 (1997)
12. Ackerman, P.J., Smalyukh, I.I.: Static three-dimensional topological solitons in fluid chiral ferromagnets and colloids. Nat. Mat. (2016). doi:10.1038/nmat4826
13. Ackerman, P.J., Smalyukh, I.I.: Diversity of knot solitons in liquid crystals manifested by linking of preimages in torons and hopfions. Phys. Rev. X **7**, 011006 (2017)

Chapter 3
Skyrmions in Chiral Magnets

This chapter covers the fundamental theories that led to the prediction of the skyrmion phase in chiral magnets. First, the Ginzburg–Landau theory of chiral magnets is shown to include an important contribution from the Dzyaloshinskii–Moriya interaction that results in a ground state with a spiral spin configuration. This spiral spin state was predicted to undergo a phase transition to the skyrmion crystal phase under the application of a magnetic field of sufficient strength. This prediction was later realized in the A-phase of MnSi and was soon found to be an ubiquitous phase in thin-film chiral ferromagnets. In addition to the standard Ginzburg–Landau analysis, modern approaches to skyrmion crystal phase such as classical Monte Carlo simulation and CP^1 theory are discussed.

3.1 Historical Survey

The formation of skyrmions in condensed matter systems has been previously documented in a number of instances, including superfluid ^{3}He and quantum Hall ferromagnets.[1] While not as widely known among the academically inclined community of condensed matter physicists, even earlier sightings of skyrmions in thin-film ferromagnets were reported in the late sixties by researchers at Bell Labs and elsewhere. The "magnetic bubble", as it was called throughout most of its lifetime, was even deemed a likely candidate to supersede the magnetic tape as a memory device, until the hard disk storage devices replaced them by virtue of their practical and tech-

[1] Volovik has written two authoritative books on superfluid ^{3}He [1, 2]. Readers who wish to familiarize themselves with topological defects in ^{3}He can start with them. The physics of skyrmions in quantum Hall ferromagnets is summarized in the edited volume by das Sarma and Pinczuk [3].

J.H. Han, *Skyrmions in Condensed Matter*, Springer Tracts in Modern Physics 278, https://doi.org/10.1007/978-3-319-69246-3_3

nical advantages.[2] From a theoretical perspective, all the examples mentioned here allowed an equal likelihood of forming skyrmions as anti-skyrmions. The "choice" of one form over the other was a matter of spontaneous symmetry breaking.

Subsequently, neutron scattering experiments in the late 70 s identified helical spin structures in noncentrosymmetric metallic ferromagnets such as MnSi and FeGe [6, 7]. The crystal structure of these materials breaks inversion symmetry,[3] and the Ginzburg-Landau (GL) theory constructed for these materials by Bak (of self-organized criticality fame) and Jensen in 1980 [8] noted the importance of Dzyaloshinskii-Moriya (DM) interaction [9–13] in understanding the resulting spiral spin structure.

Roughly a decade later, Bogdanov and collaborators developed the GL theory of chiral magnetism in the presence of an external magnetic field by including the Zeeman term in the Bak–Jensen model and looking for a saddle-point solution with a vortex-like structure. Inspired by the similarity to the Abrikosov vortex lattice in type-II superconductors, they predicted the formation of a skyrmion crystal with a triangular array in a particular magnetic field window and named a number of noncentrosymmetric materials in which the search could be conducted [14–20].

For over a decade of the new millennium, Pfleiderer's group in Germany was faithfully looking into the transport and magnetic properties of MnSi. When the "partial order" (i.e., diffuse magnetic Bragg peaks at several symmetry-related points) was discovered in 2004 [21], a number of theories [19, 22, 23] proposed a multiple spiral structure as a way to understand this feature of the neutron scattering data. These theorists understood that the multiple spiral structure could be interpreted as a crystal of topological objects such as half-skyrmions (a.k.a. merons) and hedgehogs, even though a direct experimental identification of the skyrmion lattice was still a few years away.

In MnSi, there was a curious phase, known for a long time as the A-phase, which manifested just below the Curie temperature in a rather small temperature and magnetic field window. The existence of this phase had been identified through extensive thermodynamic measurements, ranging from resistivity to magnetic susceptibility. What remained obscure for several decades, however, was the exact nature of the order parameter that set the A-phase apart. From their small-angle neutron scattering experiments, Pfleiderer's group finally concluded that this phase could be identified as the skyrmion crystal phase [24].

Yet unaware of this neutron experiment, Nagaosa and collaborators were trying to make sense of the anomalous Hall data on MnSi [25]. Their results from Monte Carlo simulations running on a discretized model of the chiral magnet had also exhibited a skyrmion in a particular magnetic field range [26]. Crucially, however, their simulation was performed in two dimensions while the neutron scattering data

---

[2] A tabloid history of the magnetic bubble memory can be found on Wikipedia. A review of the science of the bubble memory devices may be found in the book by Malozemoff and Slonczewski [4], or Eschenfelder [5], among others.

[3] Magnets that realize one specific chirality over the other will be called the chiral magnets, and the resulting spin structure as either helical or spiral spins throughout the book.

was from a bulk MnSi crystal. A year later, an experiment performed by Tokura's group on a thin film of the chiral magnet $Fe_{1-x}Co_xSi$ revealed the skyrmion phase in a wide temperature range extending down to almost zero temperature [27]. The reduced dimensionality of the thin film system provided the skyrmion lattice with greater energetic stability than in three dimensions, due to the lack of a competing conical spin phase that dominates much of the low-temperature phase diagram in three dimensions.

3.2 Ginzburg-Landau Theory of Chiral Magnets

In this chapter, we will be concerned with the theory behind the formation of skyrmion phases in materials like MnSi, an example of a crystal structure known as the B20-type. In crystallographic notation, MnSi belongs to the space group $P2_13$, and has a cubic unit cell with dimensions $a \times a \times a$, where $a = 4.558$ Å. The locations of the Mn atoms in the unit cell are given by

$$\begin{aligned} &(u, u, u)a, \quad \left(\frac{1}{2}+u, \frac{1}{2}-u, -u\right)a, \\ &\left(-u, \frac{1}{2}+u, \frac{1}{2}-u\right)a, \quad \left(\frac{1}{2}-u, -u, \frac{1}{2}+u\right)a, \end{aligned} \tag{3.1}$$

where $u_{\text{Mn}} = 0.137$. Similarly, the locations of the Si atoms in the same unit cell are given by the same formula (3.1) but with $u_{\text{Si}} = 0.845$. The unit cell size is $a = 4.558$ Å for MnSi. The crystal structure of MnSi as viewed from two different orientations is sketched in Fig. 3.1.

MnSi is a metal that undergoes a first-order ferromagnetic transition at $T_c \approx 29.5$ K. The resulting magnetic structure assumes a spiral configuration with a wavelength of $\lambda \approx 190$ Å along the [111] crystal direction, i.e., roughly the length of 23 unit cells in each orthogonal direction of the cube. The magnitude of the Mn magnetic moment deduced from a Curie–Weiss fit in the high-temperature paramagnetic phase is $2.2\mu_B$, but in the low-temperature spiral-ordered phase, this moment is reduced to $\sim 0.4\mu_B$ per Mn atom. Other metallic spiral magnets sharing the same crystal structure as MnSi include FeGe and MnGe. In FeGe, the first-order ferromagnetic transition occurs at the much higher temperature of $T \approx 278$ K, forming a spiral spin structure in the [100] direction which is then reoriented to the [111] direction at lower temperatures. The spiral wavelength is also several times larger at $\lambda \approx 700$ Å and corresponding to $\approx$150 unit cells of the FeGe crystal, for which $a = 4.7$ Å.

Among the related family of isostructural compounds, FeSi, which has one more electron per atom than MnSi, is a paramagnetic Kondo insulator. On the other hand, CoSi is a diamagnetic metal. Interestingly, the alloy compound $Fe_{1-x}Co_xSi$ becomes a magnetic metal, like MnSi or FeGe, in a substantial doping range $0.05 \leq x \leq 0.7$,

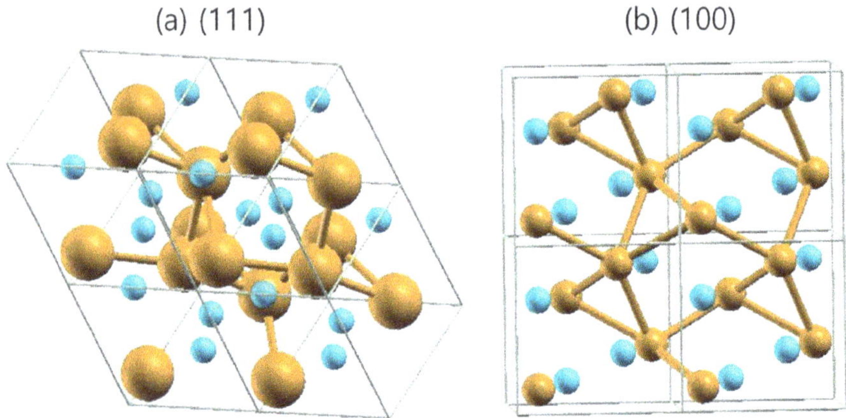

Fig. 3.1 Crystal structure of MnSi as viewed from the **a** (111) and **b** (100) crystal directions. Larger (smaller) spheres represent the Mn (Si) atoms. Figure reproduced with permission from Ref. [28]

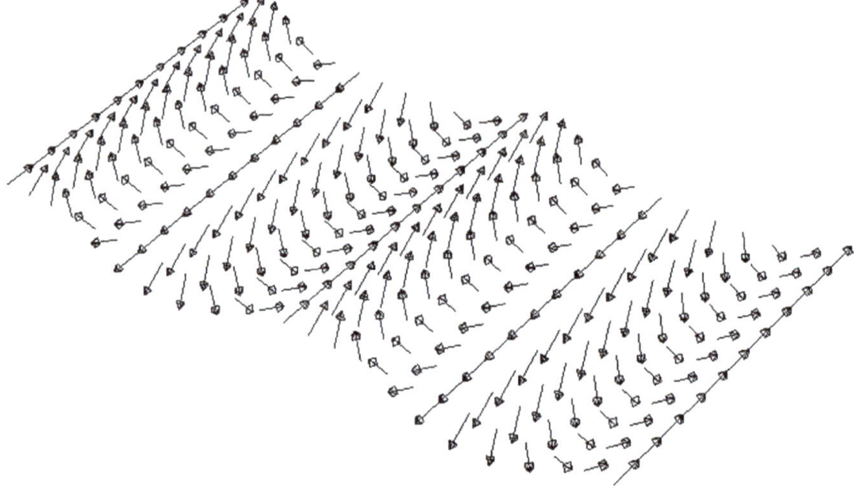

Fig. 3.2 Spiral spin structure with right-handed helicity. Chiral magnets such as MnSi exhibit this kind of spiral spin structure in the ground state. The propagation vector **k** lies along the long axis of the figure

with the spiral wavelengths ranging from 200 Å to 2000 Å along the [100] direction, depending on the doping. A summary of the spiral wavelengths in various chiral magnets can be found in the review article by Nagaosa and Tokura [29]. In all of the observed spirally ordered magnetic phases in materials mentioned thus far, the spin direction lies perpendicular to the propagation vector of the spiral. An example of such a spiral spin configuration is shown in Fig. 3.2.

A typical GL theory of the ferromagnetic moment starts with the free energy functional[4]

$$F_{\text{FM}} = \frac{J}{2} \sum_{\mu=1}^{d} (\partial_\mu \mathbf{S}) \cdot (\partial_\mu \mathbf{S}) + \frac{u}{4} (\mathbf{S}^2 - s^2)^2, \tag{3.2}$$

for a given dimensionality d. For the case of infinitely strong u and $s = 1$, this is equivalent to the classical nonlinear σ-model.

Owing to the fact that the magnetic moments are anchored at atomic sites which form a crystal structure with only discrete symmetries, and the fact that the lattice geometry affects the spin dynamics through spin-orbit coupling, some terms can arise in the free energy functional which are not invariant with respect to the continuous O(3) rotation of the spins. Taking the cubic crystal structure as an example, some typical anisotropic spin interactions are

$$F_{\text{A}} = \frac{A_1}{2} \left[\left(\frac{\partial S_x}{\partial x} \right)^2 + \left(\frac{\partial S_y}{\partial y} \right)^2 + \left(\frac{\partial S_z}{\partial z} \right)^2 \right] + A_2 (S_x^4 + S_y^4 + S_z^4). \tag{3.3}$$

A quick inspection of this functional shows that the second term is invariant only for a set of discrete rotations which are consistent with the cubic lattice symmetry, such as $(S_x, S_y, S_z) \to (S_y, S_z, S_x)$. The first term is only invariant for a simultaneous rotation of both spin and spatial coordinates in a manner that preserves the cubic symmetry. For cubic crystals, these two terms constitute the lowest-order anisotropic terms in spatial derivatives and spin components.

In cubic crystals that lack an inversion center (such as MnSi), symmetry analysis yields an additional term that may be included in the GL description, which is known as the DM interaction

$$F_{\text{DM}} = D \, \mathbf{S} \cdot (\nabla \times \mathbf{S}), \tag{3.4}$$

where the sign of D depends on the material. That F_{DM} lacks inversion symmetry is evident by the coordinate inversion, $\mathbf{r} \to -\mathbf{r}$ (implying $\nabla \to -\nabla$), which changes the sign of F_{DM}. In practice, however, inversion symmetry breaking alone is not capable of generating this term. In fact, the DM interaction is invariant under the simultaneous O(3) rotation of both spin and spatial coordinates, reflecting its origin as a spin-orbit interaction.

Some historical notes are in order. In 1958, Dzyaloshinskii published his first work on the interaction that now bears his name [9], in which he considered the appearance of uniform ferromagnetic moments in what were supposed to be pure antiferromagnets, i.e., α-Fe_2O_3, $MnCO_3$, and $CoCO_3$. Dzyaloshinskii essentially

[4]Throughout this book, theoretical models of soft magnetic moments $\mathbf{S}$ will be denoted by the free energy functional F. Theoretical models of hard magnetic moments $\mathbf{n}$ with a fixed local moment size (e.g., $\mathbf{n}^2 = 1$) will be denoted by the energy functional H.

predicted the existence of an effective interaction of the form $\mathbf{D} \cdot (\mathbf{S}_1 \times \mathbf{S}_2)$ for a pair of neighboring magnetic moments $\mathbf{S}_1$ and $\mathbf{S}_2$, where the constant vector $\mathbf{D}$ reflected the nature of the broken spatial symmetry in the atomic environment of a particular crystal. Such a term would break the symmetry under an arbitrary rotation, but would remain invariant under a restricted set of rotations that preserved the $\mathbf{D}$-vector, i.e., $\mathscr{R}\mathbf{D} = \mathbf{D}$, where $\mathscr{R}$ is a SO(3) rotation matrix. As a proof, let us note what happens if we rotate both spins with $\mathscr{R}$, i.e., $\mathbf{S}_1 \to \mathscr{R}\mathbf{S}_1$ and $\mathbf{S}_2 \to \mathscr{R}\mathbf{S}_2$. We find that

$$\mathbf{D} \cdot (\mathscr{R}\mathbf{S}_1 \times \mathscr{R}\mathbf{S}_2) = \mathscr{R}\mathbf{D} \cdot (\mathscr{R}\mathbf{S}_1 \times \mathscr{R}\mathbf{S}_2) = \mathbf{D} \cdot (\mathbf{S}_1 \times \mathbf{S}_2), \tag{3.5}$$

since $\mathscr{R}\mathbf{D} = \mathbf{D}$, and the interaction does indeed remain invariant under the restricted rotation.

Following shortly after this work, T. Moriya published his 1960 papers [12, 13], dealing with a microscopic derivation of the term Dzyaloshinskii had originally proposed out of symmetry considerations. While both Dzyaloshinskii and Moriya were interested in *antiferromagnets* containing this new term, here, we are interested in what happens to *ferromagnets* lacking inversion symmetry. Typically, microscopic derivations of the DM interaction have an insulator with a strong Hubbard-type interaction in mind, which is not directly relevant to the metallic system that interests us here. Despite these physical differences, spin-spin interactions that only arise in environments without an inversion or mirror symmetry are commonly labeled as DM interactions in the literature.

In 1964, Dzyloshinskii undertook the challenge of explaining the existence of a large-period superstructure present in certain antiferromagnets, such as MnO_2 [10]. In this work, he introduced the so-called Lifshitz invariants,

$$S_\alpha \frac{\partial S_\beta}{\partial x_\mu} - S_\beta \frac{\partial S_\alpha}{\partial x_\mu}, \quad (\mu = x, y, z) \tag{3.6}$$

for a pair of unequal spin directions α and β. He recognized that the symmetric combination $S_\alpha \partial_\mu S_\beta + S_\beta \partial_\mu S_\alpha$ only contributed a surface term to the free energy, and argued that such terms are compatible with a noncentrosymmetric crystal structure. Furthermore, he argued that these terms could be responsible for the large-period spatial modulations in antiferromagnets when combined with the Heisenberg term $\sim \sum_\mu (\partial_\mu \mathbf{S}) \cdot (\partial_\mu \mathbf{S})$ for the staggered moment.

After the discovery of large helical structures in the chiral ferromagnets MnSi and FeGe through neutron diffraction experiments, Bak and Jensen [8] proposed a GL theory consisting of all the terms mentioned in this section to explain the observed spin spirals:

$$\begin{aligned} F_{\chi\mathrm{FM}} &= \frac{J}{2} \sum_\mu (\partial_\mu \mathbf{S})^2 + \frac{u}{2} (\mathbf{S}^2 - s^2)^2 + D\mathbf{S} \cdot (\nabla \times \mathbf{S}) + F_A \\ F_A &= \frac{A_1}{2} \left[\left(\frac{\partial S_x}{\partial x} \right)^2 + \left(\frac{\partial S_y}{\partial y} \right)^2 + \left(\frac{\partial S_z}{\partial z} \right)^2 \right] + A_2 (S_x^4 + S_y^4 + S_z^4). \end{aligned} \tag{3.7}$$

We can develop a theory of the spin structure that minimizes the free energy $F_{\chi\text{FM}}$, by considering the variational spin structure

$$\mathbf{S}(\mathbf{r}) = \frac{1}{\sqrt{2}}\left(\mathbf{S}_\mathbf{k}e^{i\mathbf{k}\cdot\mathbf{r}} + \mathbf{S}_{-\mathbf{k}}e^{-i\mathbf{k}\cdot\mathbf{r}}\right), \quad \mathbf{S}_{-\mathbf{k}} = [\mathbf{S}_\mathbf{k}]^*, \tag{3.8}$$

where $\mathbf{S}_\mathbf{k}$ is some complex Fourier coefficient vector. Inserting (3.8) into the functional (3.7) and integrating over the three-dimensional space gives the free energy[5]

$$\begin{aligned}\langle F_{\chi\text{FM}} - F_\text{A}\rangle = &\frac{J}{2}\mathbf{k}^2\mathbf{S}_{-\mathbf{k}}\cdot\mathbf{S}_\mathbf{k} + iD\mathbf{S}_{-\mathbf{k}}\cdot(\mathbf{k}\times\mathbf{S}_\mathbf{k}) \\ &+ \frac{u}{4}(\mathbf{S}_{-\mathbf{k}}\cdot\mathbf{S}_\mathbf{k} - s^2)^2 + \frac{u}{8}(\mathbf{S}_\mathbf{k}\cdot\mathbf{S}_\mathbf{k})(\mathbf{S}_{-\mathbf{k}}\cdot\mathbf{S}_{-\mathbf{k}}).\end{aligned} \tag{3.9}$$

In this expression, the anisotropic term F_A is deliberately omitted in the hope that the dominant energetics are governed by the terms we have kept. Fortunately, materials such as MnSi have physical parameters that justify this assumption.

The third term on the right-hand side of (3.9) fixes the size of the Fourier component, $\mathbf{S}_{-\mathbf{k}}\cdot\mathbf{S}_\mathbf{k} = |\mathbf{S}_\mathbf{k}|^2 = s^2$, provided that $\mathbf{S}_\mathbf{k}\cdot\mathbf{S}_\mathbf{k} = 0$ is also satisfied to nullify the fourth term. We will shortly prove this to be the case. The ground state configuration minimizing (3.9) is then easy to figure out once the complex vector $\mathbf{S}_\mathbf{k}$ is decomposed into a pair of real vectors,

$$\mathbf{S}_\mathbf{k} = \boldsymbol{a}_\mathbf{k} + i\boldsymbol{b}_\mathbf{k}. \tag{3.10}$$

The DM part of the free energy then becomes

$$iD\mathbf{S}_{-\mathbf{k}}\cdot(\mathbf{k}\times\mathbf{S}_\mathbf{k}) = 2D\mathbf{k}\cdot(\mathbf{a}_\mathbf{k}\times\mathbf{b}_\mathbf{k}). \tag{3.11}$$

Given that the size of the spin $|\mathbf{S}_\mathbf{k}|^2 = [\mathbf{a}_\mathbf{k}]^2 + [\mathbf{b}_\mathbf{k}]^2$ is fixed at s^2, one can only manipulate the relative size and orientation of the $\mathbf{a}_\mathbf{k}$ and $\mathbf{b}_\mathbf{k}$ vectors in the hope of maximizing the gain from the DM energy. The most favorable condition is obtained when the $\mathbf{a}_\mathbf{k}$ and $\mathbf{b}_\mathbf{k}$ vectors are mutually orthogonal, have equal amplitudes $|\mathbf{a}_\mathbf{k}| = |\mathbf{b}_\mathbf{k}|$, and both lie in the plane normal to the $\mathbf{k}$ vector. For $D > 0$, the DM energy becomes negative if $\mathbf{k}\cdot(\mathbf{a}_\mathbf{k}\times\mathbf{b}_\mathbf{k}) < 0$. Back in terms of the real-space spin $\mathbf{S}(\mathbf{r})$, this defines a right-handed spiral. We now see that the left/right chiral symmetry is explicitly broken by the DM interaction in a chiral ferromagnet.

Taking $|\mathbf{a}_\mathbf{k}| = |\mathbf{b}_\mathbf{k}|$ and $\mathbf{k}\cdot(\mathbf{a}_\mathbf{k}\times\mathbf{b}_\mathbf{k}) < 0$, the first two terms of the free energy (3.9) become

$$J\left(|\mathbf{k}|^2 - 2\kappa|\mathbf{k}|\right)|\mathbf{a}_\mathbf{k}|^2, \quad \kappa = \frac{D}{J}. \tag{3.12}$$

[5]We are not always keeping track of the volume factor in the free energy integral. The free energy density is thus equivalent to the free energy in our system.

The spiral wavevector length is therefore fixed by the ratio of the DM energy and the exchange energy, $|\mathbf{k}| = \kappa = D/J$.[6]

Still, the orientation of the $\mathbf{k}$-vector remains to be determined. This is not surprising given the fact that, except for F_A, all the other terms in the free energy are invariant under the simultaneous O(3) rotation of spin and space. In other words, the free energy is the same as long as the triple product $\mathbf{k} \cdot (\mathbf{a_k} \times \mathbf{b_k})$ remained unchanged. This presents an interesting situation in which the ground state changes from being uniquely fixed at $\mathbf{k} = \mathbf{0}$ for $\kappa = 0$, to being a $(d-1)$-dimensional sphere of a degenerate manifold sharing the same $|\mathbf{k}| = \kappa > 0$. Lifting this degeneracy is at the mercy of the anisotropic interaction F_A.

Substituting the spin structure (3.8) into F_A turns the A_1-term into

$$\frac{A_1}{2}\left(k_x^2|S_{\mathbf{k}x}|^2 + k_y^2|S_{\mathbf{k}y}|^2 + k_z^2|S_{\mathbf{k}z}|^2\right), \tag{3.13}$$

where we recall that both $\mathbf{k}^2$ and $|\mathbf{S_k}|^2$ had their magnitudes fixed by the consideration of other free energy terms. If a given material was described by the parameter $A_1 > 0$, one would like to minimize the terms inside the parenthesis by causing each expression equal to zero. This is possible by taking $\mathbf{k} = (\kappa, 0, 0)$ and, by virtue of the orthogonality of $\mathbf{S_k}$ to the $\mathbf{k}$-vector, having $S_{\mathbf{k}x} = 0$. Then, for $A_1 > 0$, the wavevector $\mathbf{k}$ of the ground state is aligned with one of the three crystallographic axes $\hat{x}$, $\hat{y}$, or $\hat{z}$. On the other hand, for $A_1 < 0$, one would like to maximize the expression inside the parenthesis by taking $\mathbf{k}$ to lie along the [111] direction or its equivalent. For materials like MnSi, the zero-field orientation of the spiral turns out to be $\mathbf{k} \parallel [111]$, which is in good accord with the GL prediction for $A_1 < 0$.

In conclusion, the spin structure minimizing the free energy (3.7) of the chiral ferromagnet is the right-handed spiral for positive values of the DM energy $D > 0$. It is the DM term that picks one chirality of the spiral over the other, and the period λ of the spiral is also fixed by the DM energy,

$$2\pi/\lambda = \kappa = D/J. \tag{3.14}$$

Encouraged by the success of the GL approach, in the following section, we will ask what happens when the Zeeman term is added to the free energy of the chiral ferromagnet.

3.3 Skyrmion Solution in Zeeman Field

The prior symmetry consideration leading to the DM term $\mathbf{S} \cdot (\nabla \times \mathbf{S})$ in three-dimensional chiral magnets undergoes modification for thin film geometries, in which the z-direction behaves differently from the planar directions. In such thin

[6] Throughout the book, the symbol κ is reserved for the ratio $\kappa = D/J$. The sole exception is in our discussion of the CP^1 theory of chiral magnets in Sect. 3.6 where $\kappa = D/2J$ is used instead.

film geometries, the spatial anisotropy is reflected in the spin dynamics through the spin-orbit coupling, and one no longer requires the symmetry of the free energy under the full SO(3) rotation of spin and space. Instead, the allowed terms only need to be invariant under the restricted spin space rotation within the xy-plane:

$$\begin{pmatrix} S_x \\ S_y \end{pmatrix} \to \mathscr{R}_2 \begin{pmatrix} S_x \\ S_y \end{pmatrix}, \quad \begin{pmatrix} \partial_x \\ \partial_y \end{pmatrix} \to \mathscr{R}_2 \begin{pmatrix} \partial_x \\ \partial_y \end{pmatrix}, \quad \mathscr{R}_2 \in \mathrm{SO}(2). \tag{3.15}$$

Collecting all the expressions consistent with the restricted symmetry requirement yields the DM interaction appropriate for the thin-film chiral magnets:

$$\begin{aligned} H_{\mathrm{DM}} = & D_1 \left(S_x \partial_y S_z - S_y \partial_x S_z + S_z \partial_x S_y - S_z \partial_y S_x \right) \\ & + D_1' \left(S_y \partial_z S_x - S_x \partial_z S_y \right) \\ & + D_2 [(S_x \partial_x + S_y \partial_y) S_z - S_z (\partial_x S_x + \partial_y S_y)]. \end{aligned} \tag{3.16}$$

Taking $D_2 = 0$ and $D_1 = D_1' = D$ in (3.16) returns the term, $D\mathbf{S} \cdot (\nabla \times \mathbf{S})$. With the two-dimensional chiral magnets, however, the D_1'-term containing the spatial derivative ∂_z drops out, and the general DM interaction appropriate for the thin-film chiral magnet becomes

$$H_{\mathrm{DM}} = D_1 \mathbf{S} \cdot (\nabla_2 \times \mathbf{S}) + D_2 [(\mathbf{S} \cdot \nabla_2) S_z - S_z (\nabla_2 \cdot \mathbf{S})]. \tag{3.17}$$

The two-dimensional gradient defined by the shorthand $\nabla_2 = (\partial_x, \partial_y, 0)$ will be used regularly in the following discussion.

By virtue of the Derrick–Hobart theorem, we know that the nonlinear σ-model solely is not capable of supporting a skyrmion structure as a saddle-point solution since the energy is independent of the radius scale R_s. What about the DM energy $D\mathbf{S} \cdot (\nabla \times \mathbf{S})$, or its two-dimensional version (3.17)? Will this interaction be able to beat the Derrick–Hobart theorem? From dimensional analysis, the DM energy is found to scale linearly with the skyrmion size R_s:

$$D \int dxdy \, \mathbf{S} \cdot (\nabla_2 \times \mathbf{S}) \propto R_s. \tag{3.18}$$

The DM interaction either prefers $R_s = 0$ or $R_s = \infty$, depending on the sign of the integral, demonstrating that the DM interaction alone is not sufficient to stabilize the skyrmion.

The saving grace for the skyrmion finally comes from the Zeeman energy, which introduces a quadratic dependence on R_s:

$$-\int dxdy \, \mathbf{B} \cdot \mathbf{S} \propto R_s^2. \tag{3.19}$$

This R_s^2 dependence is easy to understand; over the area of the skyrmion $\propto R_s^2$, spins are generally anti-parallel to the magnetic field, running up the cost of the Zeeman energy. As a result, the Zeeman energy must be a quadratically increasing function of R_s. The energy reduction falls under the sole responsibility of the DM interaction. Intuitively, this can be understood by the tendency of the DM interaction to create spirals, and the fact that a skyrmion looks like a spiral on the local scale.

Based on the above general considerations, the overall dependence of the energy on the skymion radius R_s is expected to be of the form

$$E_s = E_0 - c_{\text{DM}} R_s + c_{\text{Z}} R_s^2, \tag{3.20}$$

where both c_{DM} and c_{Z} are positive constants. With this form, the energy minimum is likely to occur for a finite radius $R_s > 0$.

In order to establish this argument quantitatively, let us write the model skyrmion function [cf. (2.40)] in terms of the reduced variable $u = r/R_s$ and azimuthal angle φ as

$$\mathbf{S}_s = (\sin[f(u)]\cos[N\varphi], \sin[f(u)]\sin[N\varphi], \cos[f(u)]). \tag{3.21}$$

We note that the angular dependence of the $N = 1$ spin configuration is consistent with the "hedgehog-type" skyrmion shown in Fig. 3.3. With this model function, the energy of the $\mathbf{S}_s$ configuration can be calculated term-by-term.

The gradient energy of the skyrmion structure is

$$\frac{J}{2}\int dxdy \sum_{\mu=x,y} (\partial_\mu \mathbf{S}_s)\cdot(\partial_\mu \mathbf{S}_s) = \pi J \int_0^\infty du\, u\left[(f')^2 + \frac{N^2}{u^2}(\sin f)^2\right]. \tag{3.22}$$

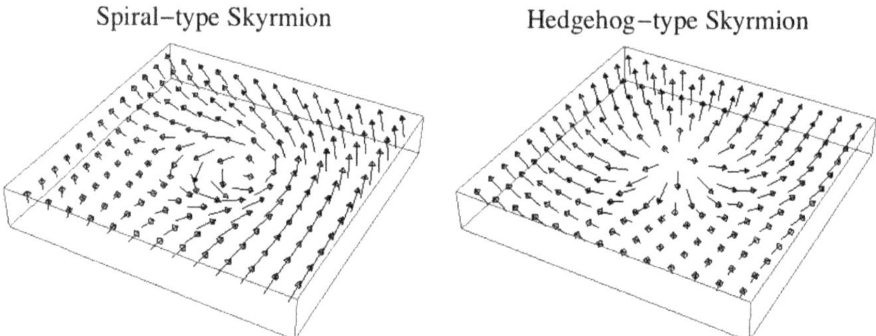

Fig. 3.3 Spiral-type (Bloch-type) skyrmion and hedgehog-type (Néel-type) skyrmion

The prime $'$ refers to the derivative with respect to the dimensionless variable $u \equiv r/R_s$. There is no R_s-dependence in this, as anticipated from our dimensional analysis.

Next we take the DM energy, considering only the first term of the general expression given in (3.17). For the skyrmion configuration (3.21), the energy vanishes:

$$H_{\rm DM} = D \int dxdy\, \mathbf{S}_s \cdot (\nabla_2 \times \mathbf{S}_s) = 0. \tag{3.23}$$

With the benefit of hindsight it is not surprising that the hedgehog-type skyrmion fails to gain energy from this type of DM interaction. The spin configuration preferred by this DM interaction is a spiral whose spins rotate in a plane orthogonal to the propagation vector. From Fig. 3.3, we can see that locally, the hedgehog skyrmion does not realize such a spin spiral structure.

A more general class of skyrmionic spin configurations carrying the same winding number N is constructed with an additional parameter φ_0:

$$\mathbf{S}_s = \Big(\sin[f(u)]\cos[N(\varphi+\varphi_0)],\, \sin[f(u)]\sin[N(\varphi+\varphi_0)],\, \cos[f(u)]\Big). \tag{3.24}$$

With this skyrmion configuration, the gradient energy $(\partial_\mu \mathbf{S})^2$ remains insensitive to φ_0, however the DM energy becomes nonzero in general:

$$\begin{aligned} H_{\rm DM} &= DR_s \left(\int_0^{2\pi} d\varphi \sin[(N-1)\varphi + N\varphi_0] \right) \\ &\quad \times \int_0^\infty du\, u \left(f' + \frac{N}{2u} \sin 2f \right). \end{aligned} \tag{3.25}$$

From the integral, we see that the only configuration which takes advantage of the DM interaction is the one with $N = +1$. Taking $N = +1$, the DM energy becomes

$$H_{\rm DM} = (2\pi D R_s \sin\varphi_0) \int_0^\infty du\, u \left(f' + \frac{\sin 2f}{2u} \right), \tag{3.26}$$

and to maximize the energy gain one must have that $\sin\varphi_0 = \pm 1$. In the earlier section we saw that $D > 0$ is consistent with the right-handed spiral being the ground state in the absence of magnetic field. Accommodating a similar spin structure within the skyrmion configuration is possible by taking $\varphi_0 = +\pi/2$, and in general, by $\varphi_0 = (\pi/2)\mathrm{sgn}(D)$. This configuration will be referred to as a spiral-type skyrmion, or a Bloch-type skyrmion.

The last term to consider is that of the Zeeman energy for the spiral-type skyrmion, which may be written as

$$H_{\rm Z} = -2\pi B R_s^2 \int_0^\infty du\, u \cos f. \tag{3.27}$$

All told, the energy of the $N = 1$ spiral-type skyrmion is given by

$$\frac{E_s}{2\pi J} = \int_0^\infty du\, u\Big(\frac{1}{2}(f')^2 + \frac{(\sin f)^2}{2u^2} + \kappa R_s\left(f' + \frac{\sin 2f}{2u}\right) + \frac{BR_s^2}{J}[1 - \cos f]\Big). \tag{3.28}$$

In this form, it looks as though the energy functional is governed by two parameters, κR_s and BR_s^2/J, but actually this is not the case. On further defining a new dimensionless variable $u' = \kappa R_s \cdot u$ and relabeling u' as the new dimensionless variable u, we retrieve the functional

$$\frac{E_s}{2\pi J} = \int_0^\infty du\, u\Big(\frac{1}{2}f'^2 + \frac{(\sin f)^2}{2u^2} + f' + \frac{\sin 2f}{2u} + \frac{B}{\kappa^2 J}[1 - \cos f]\Big), \tag{3.29}$$

with only one free parameter, $B/\kappa^2 J$. The Zeeman term here is written with an extra constant +1 to ensure that the fully polarized state $f(u) = 1$ has zero energy.

After the various considerations, we arrive at the conclusion that the "proper" dimensionless variable to describe a skyrmion in chiral magnets is not r/R_s, but $u = \kappa r$. The length scale of the skyrmion is set by κ^{-1}, the same parameter that defined the spiral wavelength through $2\pi/\lambda = \kappa$. In addition, we find a natural energy scale for the skyrmion problem to be

$$B_s = \kappa^2 J = D^2/J. \tag{3.30}$$

Now that E_s in (3.29) is a functional that depends on $f(u)$ only, we can work out its saddle-point equation from the variational calculus, $\delta E_s/\delta f = 0$:

$$f'' + \frac{f'}{u} - \frac{\sin 2f}{2u^2} + \frac{2}{u}\sin^2 f - \frac{B}{B_s}\sin f = 0. \tag{3.31}$$

This complicated nonlinear differential equation has to be solved subject to boundary conditions imposed at $u = 0$ and at $u = \infty$, for a fixed choice of the dimensionless number B/B_s. Certain qualitative features of the solution can be understood from simple reasoning. For instance, if the Zeeman field is pointing in the upward direction, $\mathbf{B} \parallel +\hat{z}$, it is natural to expect the spins to be parallel to the field at large distances from the origin, i.e., $f(\infty) = 0$. In order to obtain the skyrmion configuration, the spin at the origin must then be inverted, $f(0) = \pi$, implying that the polarity must be $P = [n^z(0) - n^z(\infty)]/2 = -1$. The skyrmion number for this situation is $Q_s = NP = -1$, which qualifies it as an anti-skyrmion. A useful analogy for the skyrmion number is to regard $Q_s = -1$ as the induced flux generated in opposition to the external flux arising from the Zeeman field. The situation then becomes analogous to the vortex nucleation in a type-II superconductor, where the superconducting current generated around the vortex flows in such a manner as to induce a flux that opposes the external one.

Solving the differential equation (3.31) subject to the boundary conditions $f(\infty) = 0,\ f(0) = \pi$ is a nice numerical task and readers are encouraged to do so on their laptops. The asymptotic solution is easy to obtain though, because one can replace $\sin f$ by f, in accordance with our assumption that $f \to 0$ at large distances. The differential equation in the asymptotic limit is given by

$$u^2 f'' + u f' - \frac{B}{B_s} u^2 f = 0 \quad (u \gg 1). \tag{3.32}$$

This is nothing but the modified Bessel equation with the solution

$$f(r) \propto K_0\left(\sqrt{\frac{B}{B_s}}u\right) = K_0\left(\sqrt{\frac{B}{J}}r\right). \tag{3.33}$$

The "healing length" $\sqrt{J/B}$ is not set by the skyrmion radius J/D, but comes from the competition between the exchange and Zeeman energies. At short distances $u \ll 1$ where $f(u) \approx \pi$, linearizing the differential equation (3.31) yields the solution

$$f(r) \approx \pi - A\kappa r \tag{3.34}$$

with an undetermined constant A.

Discussions in this section have implicitly assumed that the saddle-point solution was given by the skyrmion configuration (3.24). In fact, it is not always the case that the skyrmion structure yields the most favorable energy. For one, we know that the spiral spin structure is the ground state for $B = 0$. In contrast, for very large Zeeman fields, the only sensible ground state is the fully polarized spin state. On this basis, there must be a particular field window $B_{c1} < B < B_{c2}$, in which skyrmions can exist as stable solutions. Estimations of both the lower (B_{c1}) and upper (B_{c2}) critical fields can be performed as follows.

We start with an estimation of the skyrmion energy, based on a variational ansatz for the radial function $f(u)$ arising from the stereographic projection:

$$\cos f = \frac{u^2 - 1}{u^2 + 1}, \quad \sin f = \frac{2u}{u^2 + 1}. \tag{3.35}$$

Here, u is not equal to κr, as in the previous definition, and is rather

$$u = \kappa r / r_s, \tag{3.36}$$

with the extra variational parameter r_s that can be used to help minimize the energy. This choice of the radial function also ensures that $f' = -2/(u^2 + 1)$, and makes the energy calculation quite easy to perform. With these definitions, the skyrmion energy (3.29) is reduced to

$$\frac{E_s}{2\pi J} = \frac{2}{r_s} - 2 + \frac{B}{B_s} r_s^2 \left[\int_0^\infty \frac{dx}{1+x} \right]. \tag{3.37}$$

The log-divergent integral coming from the Zeeman energy at the far right-hand side is due to the large size of the skyrmion core region (i.e., the region where the spins are inverted) implicit in our variational function. The correct long-distance behavior of f is exponential [cf. (3.33)], while the variational form of f (3.35) has a much slower power-law decay. The divergence in the Zeeman energy can be removed by replacing the integral with a finite quantity l of order one, and then absorbing l into a redefined Zeeman field, $lB \to B$. With these approximations, one can show that the skyrmion energy (3.37) is minimized at

$$r_s = \left(\frac{B_s}{B}\right)^{1/3}, \quad \frac{E_s}{2\pi J} = 3\left(\frac{B}{B_s}\right)^{1/3} - 2. \tag{3.38}$$

The result also means that the skyrmion radius is fixed by

$$R_s = (J/D) r_s = (J/D)(B_s/B)^{1/3}. \tag{3.39}$$

Thanks to the hard work involved in the variational calculation, we could also obtain the magnetic field dependence of the skyrmion radius!

At low fields, skyrmions must compete with the spiral configuration in terms of energetics. As will be worked out in the next section, the single spiral structure has the energy density $-B_s/2$. Comparing this to E_s/A_s, where $A_s = \pi R_s^2$ stands for the area of a single skyrmion, one finds that the skyrmion energy is lower than the spiral energy when B exceeds the lower critical field B_{c1}, determined from the self-consistent equation

$$\frac{B_{c1}}{B_s} = \left(\frac{2 - B_s A_s/4\pi J}{3}\right)^3 = \left(\frac{2 - (B_s/B_{c1})^{2/3}/4}{3}\right)^3. \tag{3.40}$$

Numerically, we find that $B_{c1}/B_s \approx 0.033$.

The upper critical field value B_{c2} can be determined by noting that the ferromagnetic state has zero energy by definition, and is determined by the condition $E_s = 0$, i.e.,

$$\frac{B_{c2}}{B_s} = \left(\frac{2}{3}\right)^3 \approx 0.296. \tag{3.41}$$

The variational exercise supports the existence of an intermediate field region $B_{c1} < B < B_{c2}$, where skyrmion formation is favored over the spiral and ferromagnetic spin configurations.

The discussions in this section were based exclusively on the Bloch-type skyrmion. On the other hand, the Néel-type skyrmion (3.21) is preferred by the second term in the general DM interaction (3.17), $D[(\mathbf{S}\cdot\nabla_2)S_z - S_z(\nabla_2\cdot\mathbf{S})]$. Using this form of the DM energy, one can work out the differential equation for the radial function to obtain the hedgehog-type skyrmion profile. This was completed early on by Bogdanov and Rößler [17], and was revisited recently by Rohart and Thiaville [30]. This kind of DM interaction plays a prominent role in the heavy metal/transition metal/non-magnetic metal multilayer structure.

3.4 Topological Crystals as a Multiple Spiral Phase

The GL theory for chiral magnets correctly predicted the spiral spin ground state, and when the Zeeman field was included, the same theory produced the skyrmion solution at the saddle point. The energetic stability of one nucleated skyrmion implies that many others can form, eventually covering the whole lattice and forming a skyrmion crystal. In the previous section we saw that describing one skyrmion with mathematical rigor was a fairly elaborate process. It turns out that understanding the skyrmion lattice is in some sense simpler, as it can be described as a superposition of a number of spirals. The idea of multiple spiral state proves to be more general than for just the two-dimensional skyrmion lattice. As we will see, the three-dimensional multi-spiral state can be interpreted as a topological crystal made up of hedgehog and anti-hedgehog defects.

A multi-spiral state refers to a spin configuration given by

$$\mathbf{S}(\mathbf{r}) = \mathbf{S}_0 + \sum_{\alpha=1}^{N}\left(\mathbf{S}_\alpha e^{i\mathbf{k}_\alpha\cdot\mathbf{r}} + \mathbf{S}_\alpha^* e^{-i\mathbf{k}_\alpha\cdot\mathbf{r}}\right). \tag{3.42}$$

Each α refers to a spin spiral with a definite wavevector $\mathbf{k}_\alpha$, whose spins rotate in a plane orthogonal to $\mathbf{k}_\alpha$. The additional uniform magnetization term $\mathbf{S}_0 = (0, 0, n_0)$ is introduced in the variational spin configuration, in anticipation of the effects of a spin-polarizing Zeeman field directed along the $+\hat{z}$ direction.

The issue we must face is what type of multi-spiral structure is favored by the generic GL energy functional of the chiral magnet [cf. (3.7)]

$$\begin{aligned} F &= \frac{J}{2}\sum_\mu\left(\partial_\mu\mathbf{S} - \kappa\hat{e}_\mu\times\mathbf{S}\right)^2 + \frac{u}{4}(\mathbf{S}^2-1)^2 - \mathbf{B}\cdot\mathbf{S} \\ &+ \frac{A_1}{2}\left[\left(\frac{\partial S_x}{\partial x}\right)^2 + \left(\frac{\partial S_y}{\partial y}\right)^2 + \left(\frac{\partial S_z}{\partial z}\right)^2\right] + A_2(S_x^4+S_y^4+S_z^4). \end{aligned} \tag{3.43}$$

The unusual expression in the first term follows from a piece of mathematical trickery that combines the exchange and DM interactions:

$$\frac{J}{2}\sum_\mu(\partial_\mu\mathbf{S})^2 + D\mathbf{S}\cdot(\nabla\times\mathbf{S}) = \frac{J}{2}\sum_\mu(\partial_\mu - \kappa\hat{e}_\mu\times\mathbf{S})^2 - \frac{D^2}{2J}. \tag{3.44}$$

This identity holds for three-dimensional space where $\mu = x, y, z$. Truncating ∂_z and $\hat{e}_z$ from both sides of the equation yields a valid expression in two dimensions. Since $D > 0$ favors a right-handed spiral, we will assume that all the spirals in the multi-spiral configuration (3.42) are right-handed and have the same amplitude:

$$\mathbf{S}_\alpha = \frac{n_s}{2} e^{i\theta_\alpha} (\hat{e}_{1\alpha} - i\hat{e}_{2\alpha}). \tag{3.45}$$

Any pair of orthonormal vectors forming the triad, $\hat{e}_{1\alpha} \times \hat{e}_{2\alpha} = \hat{k}_\alpha$, qualify as basis vectors. An arbitrary phase θ_α is assigned for each spiral mode for greater generality. All the spiral wave vectors $\mathbf{k}_\alpha = k\hat{k}_\alpha$ are taken to be of equal length. The multi-spiral configuration generally does not have a uniform size of the moment $\mathbf{S}$ over space, and the best one can do to accommodate the u-term in the free energy is to impose the average constraint

$$\langle \mathbf{S}(\mathbf{r}) \cdot \mathbf{S}(\mathbf{r}) \rangle = n_0^2 + N n_s^2 = 1. \tag{3.46}$$

At first sight, feeding the multi-spiral ansatz (3.42) into the free energy (3.43) and carrying out the minimization seems a rather formidable task. However, suppose that we could remove the u-interaction from the free energy, since the constraint has already been met in (3.46), and the two anisotropy terms A_1 and A_2 can be ignored to zeroth order against the more pressing exchange, DM, and Zeeman interactions. Under these simplifications, the multi-spiral ansatz (3.42) yields the surprisingly simple energy form,

$$E = \frac{1}{2} N n_s^2 \left(J k^2 - 2Dk \right) - B n_0. \tag{3.47}$$

Minimizing this energy with respect to k by solving $\partial E / \partial k = 0$ gives the expected spiral wavelength relation $k = 2\pi/\lambda = \kappa$, and the energy

$$E = -\frac{D^2}{2J} N n_s^2 - B n_0 = \frac{D^2}{2J} n_0^2 - B n_0 - \frac{D^2}{2J}. \tag{3.48}$$

This is a function with respect to the ferromagnetic moment n_0, which is yet to be minimized. If we did that, we would find

$$n_0 = \frac{B}{D^2/J} = \frac{B}{B_s}, \quad E = -\frac{1}{2} \frac{B^2}{B_s} - \frac{1}{2} B_s. \tag{3.49}$$

The energy scale $B_s = D^2/J$ we encountered in the skyrmion calculation of Sect. 3.3 appears here as well.

The most interesting outcome of this simple analysis is that the energy remains completely insensitive to the number of spirals being superposed, N, and to the orientation of each spiral $\hat{k}_\alpha$. This represents a massive degeneracy in the multi-spiral solution, which can only be lifted by the residual u, A_1, and A_2 terms in the free energy. This observation suggests an obvious strategy: a variational calculation comparing the energies of several candidate multi-spiral states can be performed in the (u, A_1, A_2) parameter space to identify the true ground state. A number of phase diagrams for chiral magnets have been proposed on the basis of GL free energy analysis of multi-spiral states [8, 19, 22, 23, 31]. Nowadays, a more popular alternative method to determine the phase diagram of the chiral magnets is to run a Monte Carlo simulation on a lattice version of the Hamiltonian.

Still, there are several valuable lessons one can learn from viewing the multi-spiral state as a natural way of generating topological crystal structures in two and three dimensions. In the following two subsections, we will discuss this aspect in detail for each dimension. The alternative Monte Carlo method will be the subject of Sect. 3.5.

3.4.1 *Two-Dimensional Multi-spiral States*

Several variational multi-spiral states can be considered in two dimensions, e.g., single-spiral, double-spiral, and triple-spiral states. As suggested in the previous section, each of these states is characterized by specifying the orientations of the spirals.

The single-spiral state has the wave vector $\mathbf{k} = \kappa(1, 1, 0)/\sqrt{2}$. This state forms the ground state in the region of small magnetic field. The double-spiral state with two orthogonal wavevectors $\mathbf{k}_1 = \kappa(1, 0, 0)$ and $\mathbf{k}_2 = \kappa(0, 1, 0)$ gives rise to a spin configuration that is equivalent to a square lattice of alternating half-skyrmions and half-anti-skyrmions, as first shown in Ref. [19]. The triple-spiral state with three wavevectors at a relative angle of 120° to one another, $\mathbf{k}_1 = \kappa(1, 0, 0)$, $\mathbf{k}_2 = \kappa\left(-1, \sqrt{3}, 0\right)/2$, and $\mathbf{k}_3 = \kappa\left(-1, -\sqrt{3}, 0\right)/2$. is equivalent to a triangular lattice of skyrmions such as that found in the A-phase of MnSi and in a 2D thin-film samples.

Feeding each variational state into the free energy functional (3.43), one obtains an energy that depends on (u, A_1, A_2, B) and n_0. By first minimizing the energy with respect to n_0, one obtains the energy that only depends on (u, A_1, A_2, B). The energies of the various candidate states can then be compared for a given parameter value. The task, while straightforward, is quite involved and readers interested in the details of the calculation and the ensuing phase diagram may refer to Refs. [22, 23, 31] for further details.

A more instructive exercise is to try to understand why the multi-spiral phase is compatible with the notion of a topological spin crystal. Let us take the double-spiral phase as an example,

$$\mathbf{n} = (\sin y, \cos x, \cos y + \sin x + m). \tag{3.50}$$

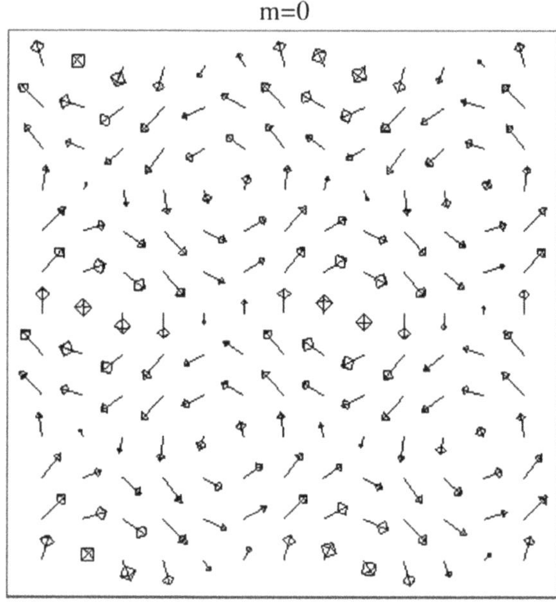

Fig. 3.4 Square lattice of skyrmions constructed using (3.50) with $m = 0$

The overall normalization factor has been ignored since it is irrelevant in the calculation of topological density, and we have chosen $k \equiv 1$ to keep the notation as simple as possible. A useful way to analyze the topological structure of the spin configuration is to first identify the nodal positions. These are the points in space where $\mathbf{n}$ vanishes. For $m = 0$, the nodal points occur at $(x_0, y_0) = (-\pi/2, 0)$ and $(x_0, y_0) = (\pi/2, \pi)$ in a unit cell of size $(2\pi)^2$ (Fig. 3.4).

Expansion of $\mathbf{n}(x, y)$ around one such node (x_0, y_0) yields the linearized spin structure

$$\mathbf{n} \propto (y \cos y_0, -x \sin x_0, m - y \cos y_0 + x \sin x_0), \tag{3.51}$$

where the coordinates (x, y) are measured with respect to the nodal position. The linearized spin profile gives rise to the skyrmion density

$$\begin{aligned} \rho(\mathbf{r}) &= \frac{1}{4\pi|\mathbf{n}|^3}\mathbf{n} \cdot \left(\frac{\partial \mathbf{n}}{\partial x} \times \frac{\partial \mathbf{n}}{\partial y}\right) = \frac{1}{4\pi}\frac{P(m)}{[X^2 + Y^2 + P(m)^2]^{3/2}} \\ P(m) &= m\frac{\sin x_0 \cos y_0}{\sqrt{1 - \cos^2 x_0 \sin^2 y_0}}. \end{aligned} \tag{3.52}$$

The capitalized coordinates (X, Y) are some orthogonal rotation of (x, y), and for both nodal positions one finds $P(m) = -m$. Integrating $\rho(X, Y)$ over the (X, Y) space gives the skyrmion number

$$Q(m) = -\frac{1}{2}\mathrm{sgn}(m). \tag{3.53}$$

This formula can be interpreted as follows. As m passes from positive to negative values, the skyrmion number changes from $-1/2$ to $+1/2$. With two such nodes per unit cell, the total skyrmion number per unit cell depends on m as $-\mathrm{sgn}(m)$.

It is instructive to view the change of the skyrmion number in the enlarged three-dimensional space consisting of two spatial coordinates (x, y) and one parameter m. Each three-dimensional coordinates $(x, y, m) = (x_0, y_0, 0)$, where (x_0, y_0) is the nodal position for $m = 0$, defines the center of a hedgehog or an anti-hedgehog. A passage through the hedgehog center in the m-direction changes the skyrmion number by one.

In the case of the triple-spiral state, the spin structure

$$\mathbf{n}=\left(\sqrt{3}\cos\frac{x}{2}\sin\frac{\sqrt{3}y}{2}, -\left(2\cos\frac{x}{2}+\cos\frac{\sqrt{3}y}{2}\right)\sin\frac{x}{2}, \cos x+2\cos\frac{x}{2}\cos\frac{\sqrt{3}y}{2}\right)+m \tag{3.54}$$

has no node in the entire space of (x, y), even for $m = 0$. One finds $Q = -1$ by direct integration of the skyrmion density over one unit cell of the triangular lattice, irrespective of the value of m (as long as it is not too large). This state corresponds to the phase crystallized in thin-film chiral magnets subject to a perpendicular magnetic field of moderate strength (Fig. 3.5).

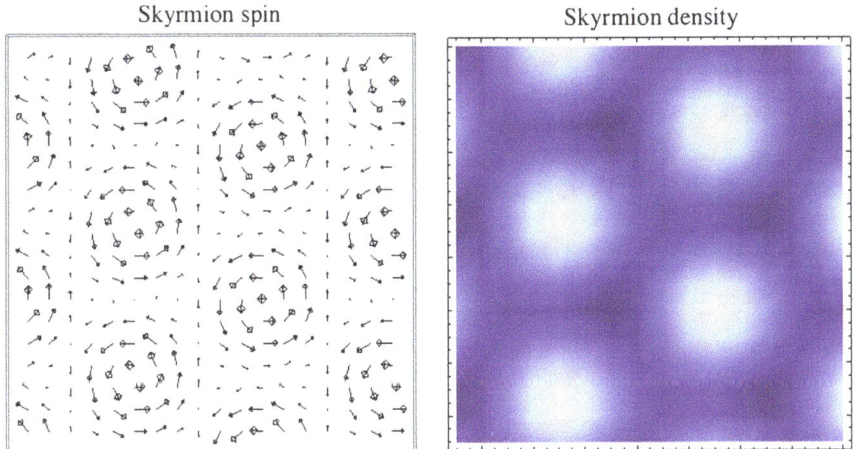

Fig. 3.5 A triangular skyrmion lattice generated by the function (3.54) with $m = 0$. In the skyrmion density plot on the right, regions with large negative values are indicated in white

3.4.2 Three-Dimensional Multi-spiral States

A similar calculation strategy applies for three-dimensional multi-spiral spins. Some of the candidate states are mere translations of the two-dimensional structures into the third direction, such as the columnar skyrmion lattice obtained from uniform translation of the triple-spiral state in the z-direction. Other structures are generic to three dimensions.

The single-spiral phase with $\mathbf{k} = k(1, 1, 1)/\sqrt{3}$ forms the ground state at low magnetic fields ($B < B_{c1}$) in bulk MnSi crystal. At low temperatures and for intermediate magnetic fields one finds the conical state with $\mathbf{k} = k(0, 0, 1)$, which turns into the ferromagnetic state as the field gets stronger.

A truly three-dimensional topological crystal lattice is realized by superposing three orthogonal spin spirals: $\mathbf{k}_1 = k(1, 0, 0)$, $\mathbf{k}_2 = k(0, 1, 0)$ and $\mathbf{k}_3 = k(0, 0, 1)$. The resulting configuration is a realization of the simple-cubic lattice of hedgehogs and anti-hedgehogs. Other structures like face-centered-cubic and body-centered cubic crystals of hedgehogs can be formed by the superposition of four (FCC) or six (BCC) spirals. Details of their construction and the phase diagrams can be found in Refs. [22, 23, 31].

Analysis of the topological structure for three-dimensional crystals can be conveniently done by defining the layer-dependent topological number

$$Q(z) = \int dxdy\, \rho(x, y, z). \tag{3.55}$$

Let us apply the scheme to the triple-spiral state as an example:

$$\mathbf{n} = (\sin y + \cos z, \sin z + \cos x, \sin x + \cos y + m). \tag{3.56}$$

Again, the overall normalization is ignored. There are eight independent zeroes (nodes) in a unit cell of size $(2\pi)^3$ for $m = 0$. Considering the Taylor-expansion around one such nodal position (x_0, y_0, z_0) to first order in the deviation results in

$$\mathbf{n} = (y \cos y_0 - z \sin z_0, z \cos z_0 - x \sin x_0, x \cos x_0 - y \sin y_0). \tag{3.57}$$

As in the two-dimensional case, the coordinates (x, y, z) are measured from their nodal positions. The skyrmion density for this configuration is

$$\begin{aligned} \rho(x, y, z) &= \frac{1}{4\pi} \frac{P(z, m)}{[X^2 + Y^2 + P(z, m)^2]^{3/2}}, \\ P(z, m) &= z \frac{\cos x_0 \cos y_0 \cos z_0 - \sin x_0 \sin y_0 \sin z_0}{\sqrt{1 - \cos^2 x_0 \sin^2 y_0}}, \end{aligned} \tag{3.58}$$

and (X, Y) is again related to (x, y) by an orthogonal coordinate change. Integrating over the entire (X, Y) space yields the layer-dependent skyrmion number

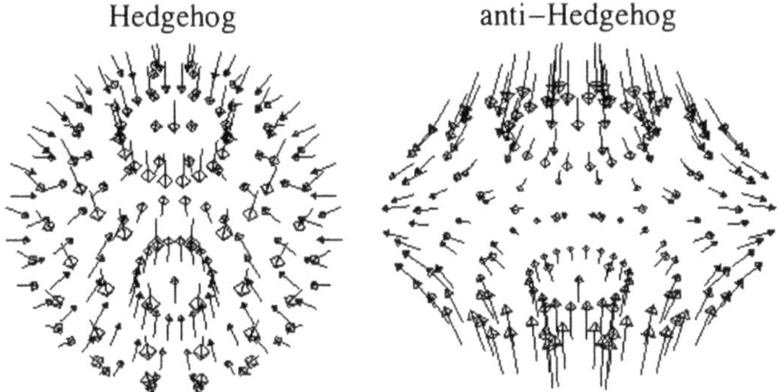

Fig. 3.6 Hedgehog and anti-hedgehog spin configurations within the simple-cubic spin crystal. The magnetization vector **n** vanishes at the center of each singularity

$$Q(z) = \frac{1}{2}\text{sgn}[P(z,m)]. \tag{3.59}$$

Since $P(z, m)$ is proportional to z, one finds that $Q(z)$ jumps by +1 or -1 as the nodal position is crossed along the z-direction. We have either a hedgehog or an anti-hedgehog centered at (x_0, y_0, z_0). It turns out that among the eight nodes in a unit cell there are an equal number of hedgehogs as there are anti-hedghogs. A snapshot of the hedgehog and anti-hedgehog configurations obtained from (3.57) is shown in Fig. 3.6.

In two dimensions, it was possible to construct a singularity-free topological lattice as a superposition of three spirals. In three dimensions, we showed how to construct a topological crystal with a lattice of nodes, or hedgehog centers, by superposing three or more spirals. A nodeless, three-dimensional topological crystal would be a Skyrme crystal in the veritable sense of the word, but is much harder to construct. The reason behind the difficulty for creating a singularity-free structure in three dimensions is clear. Each component of the spin vector that is taken equal to zero defines a two-dimensional surface embedded in a three-dimensional space. For two-dimensional spin structures, such a surface is extended vertically along the z-axis and traces out a curve in the xy-plane. The intersection of three curves, from each zero-component of the spin vector, defines the node. The three curves in a plane do not cross at a single point except through fine-tuning, such as the tuning of m through zero in the double-spiral state. For both double and triple spirals, the nodeless spin structures also happen to be topologically nontrivial and carry a finite skyrmion number. The same reasoning applied to a generic three-dimensional spin crystal structure states that nodes are generically unavoidable because three independent surfaces do cross at isolated points.

3.5 Monte Carlo Simulations

Classical Monte Carlo simulations have played an exceptionally important role in establishing the phase diagrams of chiral magnets in both two and three dimensions. The match between the simulation-generated phase diagrams and experimental diagrams is usually very good. Such agreement could be anticipated, after all, since we know that Monte Carlo simulation, being essentially an exact scheme, gives the most accurate phase diagram of the model for arbitrary temperatures and parameters. The scheme is very simple: one first converts the continuum free energy of the chiral magnet to a suitably discretized lattice model, and then runs a single spin-flip Monte Carlo algorithm. In this section, we discuss how to generate such lattice models in two and three dimensions and consider the simulation results that follow.

3.5.1 Two Dimensional Simulations

The Heisenberg exchange part $(J/2)\sum_\mu (\partial_\mu \mathbf{S})^2$ of the GL free energy can be derived as the continuum limit of the lattice Hamiltonian

$$H_{\mathrm{H}} = -J \sum_{i,\mu} \mathbf{n}_i \cdot \mathbf{n}_{i+a\hat{e}_\mu}, \tag{3.60}$$

where $a\hat{e}$ is a vector connecting one lattice site i to its neighboring sites at $i + a\hat{e}_\mu$ on a square ($\mu = x, y$) or cubic ($\mu = x, y, z$) lattice. A simple Taylor expansion of (3.60), assuming a small lattice constant a, gives the first term $\mathbf{n}_i \cdot \mathbf{n}_i$ that is equal to unity and can be discarded as a constant term. The second term $\sum_i \mathbf{n}_i \cdot \partial_\mu \mathbf{n}_i$ is a total derivative and can be similarly discarded. Finally, the third term $(a^2/2)\sum_i \mathbf{n}_i \cdot \partial_\mu^2 \mathbf{n}_i$ leads, in the continuum limit, to

$$H_{\mathrm{H}} \approx -\frac{1}{2} J a^2 \sum_i \mathbf{n}_i \cdot \partial_\mu^2 \mathbf{n}_i \to \frac{1}{2} J a^{2-d} \int d^d\mathbf{r}\, (\partial_\mu \mathbf{n}) \cdot (\partial_\mu \mathbf{n}). \tag{3.61}$$

With a proper re-definition of the parameters, one recovers the $(J/2)(\partial_\mu \mathbf{S}) \cdot (\partial_\mu \mathbf{S})$ term of the continuum GL theory.

How we can recover the continuum DM interaction from the lattice Hamiltonian is also not hard to figure out. Unlike the Heisenberg exchange $\mathbf{S}_i \cdot \mathbf{S}_j$ which is symmetric under the exchange of two spins $\mathbf{S}_i \leftrightarrow \mathbf{S}_j$, there can also be an antisymmetric exchange of spins $\mathbf{S}_i \times \mathbf{S}_j$ which changes sign. To form a scalar from the antisymmetric exchange, a third vector needs to be introduced to make an inner product. The most obvious candidate for each $\langle ij \rangle$ pair is the vector connecting the two sites. Our first guess for the DM interaction at the lattice level is

$$H_{\text{DM}} = -D \sum_{i,\mu} \hat{e}_\mu \cdot (\mathbf{n}_i \times \mathbf{n}_{i+a\hat{e}_\mu}). \tag{3.62}$$

Upon Taylor expansion, the first derivative of $\mathbf{n}_{i+a\hat{e}_\mu}$ provides the contribution

$$H_{\text{DM}} \sim Da \sum_{i,\mu} \hat{e}_\mu \cdot (\partial_\mu \mathbf{n}_i \times \mathbf{n}_i) = Da \sum_{i,\mu} \mathbf{n}_i \cdot (\hat{e}_\mu \times \partial_\mu \mathbf{n}_i). \tag{3.63}$$

A short exercise in vector calculus shows that $\sum_\mu \hat{e}_\mu \times \partial_\mu \mathbf{n} = \nabla \times \mathbf{n}$, yielding

$$H_{\text{DM}} = Da^{1-d} \int d^d\mathbf{r}\, \mathbf{n} \cdot (\nabla \times \mathbf{n}), \tag{3.64}$$

in the continuum limit. The GL result is recovered with the rescaling $Da^{1-d} \to D$. The lattice analogue of the continuum GL theory of the chiral magnet thus reads

$$H_{\text{HDM}} = -J \sum_{i,\mu} \hat{e}_\mu \cdot (\mathbf{n}_i \times \mathbf{n}_{i+a\hat{e}_\mu}) - D \sum_{i,\mu} \hat{e}_\mu \cdot (\mathbf{n}_i \times \mathbf{n}_{i+a\hat{e}_\mu}). \tag{3.65}$$

The acronym HDM is for Heisenberg, Dzyaloshinskii, and Moriya, three venerable names in the theory of magnetism. We can extend this model by adding a fourth name associated with the Zeeman energy

$$H_Z = -\mathbf{B} \cdot \sum_i \mathbf{n}_i. \tag{3.66}$$

Experience has proven the HDMZ Hamiltonian,

$$\begin{aligned} H_{\text{HDMZ}} =& H_{\text{HDM}} + H_Z \\ =& -J \sum_{i,\mu} \hat{e}_\mu \cdot (\mathbf{n}_i \times \mathbf{n}_{i+a\hat{e}_\mu}) - D \sum_{i,\mu} \hat{e}_\mu \cdot (\mathbf{n}_i \times \mathbf{n}_{i+a\hat{e}_\mu}) \\ & - \mathbf{B} \cdot \sum_i \mathbf{n}_i, \end{aligned} \tag{3.67}$$

to be the minimal lattice model, and that it is sufficient to capture all the essential features of the field-driven skyrmion lattice formation in two dimensions.

The form of the DM interaction used in the above equation favors the formation of spiral-type skyrmions. If one wants to realize the hedgehog-type skyrmion instead, one should use the lattice version of the second term in the general two-dimensional DM energy, (3.17),

$$-D \sum_i [\hat{y} \cdot (\mathbf{n}_i \times \mathbf{n}_{i+\hat{x}}) - \hat{x} \cdot (\mathbf{n}_i \times \mathbf{n}_{i+\hat{y}})]. \tag{3.68}$$

As noted earlier, DM interactions of this kind can be realized in an interfacial geometry with broken inversion symmetry between the $z > 0$ and $z < 0$ regions. One can even conceive a general two-dimensional lattice DM model that mixes both interactions,

$$-D\sum_i\Big[\hat{d}_1\cdot(\mathbf{n}_i\times\mathbf{n}_{i+\hat{x}})+\hat{d}_2\cdot(\mathbf{n}_i\times\mathbf{n}_{i+\hat{y}})\Big], \tag{3.69}$$

where the two orthogonal unit vectors are defined by

$$\begin{aligned}\hat{d}_1 &= \hat{x}\cos b+\hat{y}\sin b,\\ \hat{d}_2 &= \hat{y}\cos b-\hat{x}\sin b,\end{aligned} \tag{3.70}$$

with some angle b. Thus far, typical Monte Carlo calculations have focused only on $b = 0$ or $b = \pi/2$ limits. Finally, one could further embellish the lattice Hamiltonian with various anisotropy terms

$$H_{\rm A} = A_1\sum_i\Big[(n_i^x)^4+(n_i^y)^4+(n_i^z)^4\Big]-A_2\sum_i\Big[n_i^x n_{i+\hat{x}}^x+n_i^y n_{i+\hat{y}}^y\Big]. \tag{3.71}$$

Nowadays, numerous Monte Carlo (MC) simulations of the skyrmion lattice have been performed on the basis of the HDMZ model and its variants. One can easily locate some examples by googling the term “skyrmion lattice”. One early example of a MC-generated spin configuration is shown in Fig. 3.9b. Indeed, generating skyrmion lattice configurations by MC simulation is a fun (and easy!) exercise for those who wish to learn about MC techniques and how to apply them.

Once the MC program is ready, you start the run at a high temperature several times J, and gradually lower the temperature T in intervals of a small fraction of J down to $T\sim 0$. Unless one intends to study thermodynamic quantities such as the specific heat and magnetic susceptibility at each temperature, or to probe critical properties near the ferromagnetic phase transition, there is no need to perform a large number of MC runs in the high-temperature region. Roughly 5×10^4 runs at each T are sufficient to reach the correct ground state for a typical $L\times L$ lattice. Although, for large simulations with $L\sim 100$, the run might have to be longer to ensure good convergence. Since the skyrmion lattice is a crystal phase of sort, there is the usual concern regarding the formation of dislocations, interstitials, and so forth in the final state of the MC run. In practice, this sort of problem does not occur provided the annealing procedure is carried out slowly enough.

A problem of practical interest in carrying out the MC calculation is the determination of the spiral length, which is not given by the simple ratio $2\pi/\lambda = J/D$ in the lattice model. For $\mathbf{B} = 0$, the minimum energy of the HDM Hamiltonian for a single-spiral with $\mathbf{k} = k(1, 1)$ is found at

$$D/J = \sqrt{2}\tan k. \tag{3.72}$$

To realize a spiral with a period of 6 lattice sites, for example, one must choose $D/J = \sqrt{2}\tan(2\pi/6) = \sqrt{6}$, so actually D has to be much larger than J in the simulation. In typical MC simulations, the lattice size L is made commensurate with the spiral period, and the skyrmion lattice usually forms under some finite **B**-field with the inter-skyrmion spacing also given by the same period. Once the value of D is fixed, one can run the MC calculation for increasing value of B to observe the evolution of the phase from the (1,1)-spiral to the skyrmion lattice, and then to the ferromagnetic state.

3.5.2 *Three-Dimensional Simulations*

Unlike the two-dimensional chiral magnet that was experimentally investigated relatively recently, the phase diagram of the three-dimensional chiral magnet has been known for decades. Its most mysterious feature, the A-phase, was later proven to host the columnar skyrmion lattice. The first successful demonstration of the A-phase in a three-dimensional model of a chiral magnet by MC simulation was achieved by Buhrandt and Fritz [32] using the model Hamiltonian

$$\begin{aligned} H_{\mathrm{BF}} = & -J\sum_{\mathbf{r}} \mathbf{n_r}\cdot(\mathbf{n}_{\mathbf{r}+\hat{x}} + \mathbf{n}_{\mathbf{r}+\hat{y}} + \mathbf{n}_{\mathbf{r}+\hat{z}}) \\ & + \frac{J}{16}\mathbf{n_r}\cdot(\mathbf{n}_{\mathbf{r}+2\hat{x}} + \mathbf{n}_{\mathbf{r}+2\hat{y}} + \mathbf{n}_{\mathbf{r}+2\hat{z}}) \\ & - D\sum_{\mathbf{r}} \mathbf{n_r}\cdot(\mathbf{n}_{\mathbf{r}+\hat{x}}\times\hat{x} + \mathbf{n}_{\mathbf{r}+\hat{y}}\times\hat{y} + \mathbf{n}_{\mathbf{r}+\hat{z}}\times\hat{z}) \\ & + \frac{D}{8}\sum_{\mathbf{r}} \mathbf{n_r}\cdot(\mathbf{n}_{\mathbf{r}+2\hat{x}}\times\hat{x} + \mathbf{n}_{\mathbf{r}+2\hat{y}}\times\hat{y} + \mathbf{n}_{\mathbf{r}+2\hat{z}}\times\hat{z}) \\ & - \mathbf{B}\cdot\sum_{\mathbf{r}} \mathbf{n_r}. \end{aligned} \tag{3.73}$$

Their model differs from the straightforward 3D generalization of the HDMZ Hamiltonian in that two next-nearest-neighbor interactions have been introduced with specific choices of the constants $J/16$ and $D/8$. Taylor expansion of these terms shows that they cancel the fourth-order anisotropy terms coming from the nearest-neighbor J and D terms, thereby producing a less anisotropic model than the original HDMZ Hamiltonian. The weakened anisotropy effect helps to stabilize the finite temperature A-phase in the MC simulation. The phase diagram they produced is reproduced as Fig. 3.7, and looks very similar to the one that experimentalists have known for decades. We see that the skyrmion crystal phase that took up most of the low-temperature phase diagram in two dimensions is gradually pushed up to higher temperature regimes as the lattice thickness in the z-direction is increased, until it is squeezed into the A-phase pocket of the fully three-dimensional model.

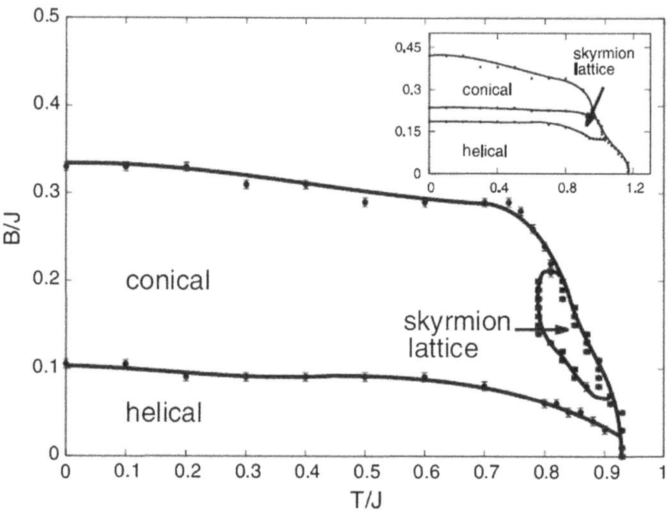

Fig. 3.7 Phase diagram of the three-dimensional Hamiltonian for chiral magnetic materials. Overall, the diagram shows good qualitative agreement with the experimentally determined phase diagram of MnSi and other chiral magnets. The inset diagram is obtained when next-nearest-neighbor interactions are excluded in (3.73). Reproduced with permission from Ref. [32]

A few years after the Buhrandt-Fritz work, the same model Hamiltonian was studied for increasing D/J values [33]. The resulting MC phase diagram showed that the A-phase could extend down to the lower temperature regime, and above some critical ratio $(D/J)_c$, eventually turn into a simple cubic lattice of monopoles (hedgehogs) and anti-monopoles (anti-hedgehogs). An example of the monopole-anti-monopole lattice configuration is shown in Fig. 3.8. This was one of the three-dimensional topological crystal phases predicted to occur by the GL analysis. A larger D/J value implies the shorter wavelength spiral. For instance, the spiral wavelength of the [111] spiral $\mathbf{k} = k(1, 1, 1)/\sqrt{3}$ is given by the simple formula

$$\frac{D}{J} = \frac{\sqrt{3}\sin k(4 - \cos k)}{4\cos k - \cos 2k}. \tag{3.74}$$

The MC calculation can be run on a cubic lattice with periodic boundary conditions, using values of D/J chosen to give the integer spiral period $\lambda = 2\pi/q$. A further technical improvement made in Ref. [33] was the use of GPU chips that carried out spin flip operations over hundreds of processors at once. While each processing unit had a slower speed than a single CPU, the sheer number of operating units in a single GPU chip allowed superior efficiency in the running time of MC calculation over a standard CPU-based simulation. In an encouraging note, there are strong experimental indications that three-dimensional topological spin structure exists in the short-wavelength spiral magnet MnGe [34].

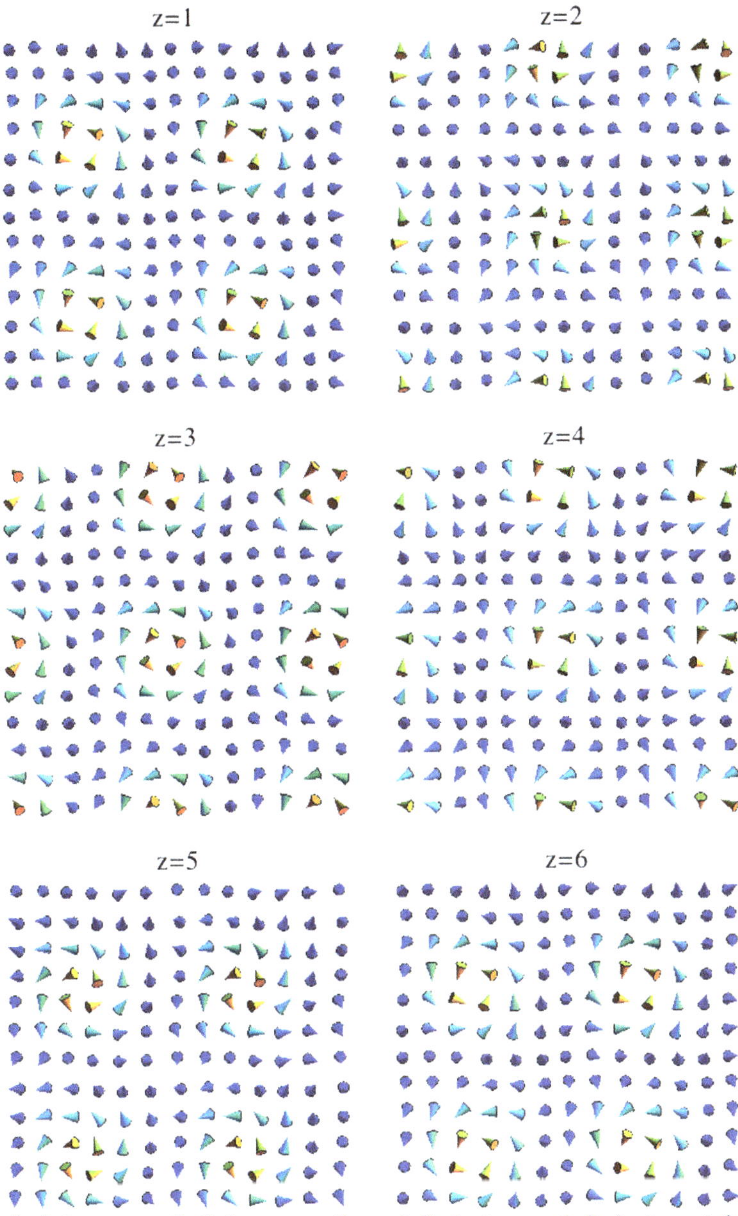

Fig. 3.8 Layer-by-layer spin configuration of the hedgehog crystal obtained from Monte Carlo simulation of the BF Hamiltonian (3.73) with $\lambda = 6$. Calculation and figure preparation done by Seong-Gyu Yang

3.6 CP¹ Theory of the Skyrmion Crystal

In Chap. 2 [cf. (2.71)], we saw how the nonlinear σ-model for a ferromagnet had the equivalent CP¹ formulation:

$$\frac{J}{2}\sum_{\mu}(\partial_\mu \mathbf{n})\cdot(\partial_\mu \mathbf{n}) = 2J\sum_{\mu}(D_\mu \mathbf{z})^\dagger (D_\mu \mathbf{z}). \tag{3.75}$$

The covariant derivative $D_\mu = \partial_\mu - ia_\mu$ is defined in terms of the abelian gauge field $a_\mu = -i\mathbf{z}^\dagger \partial_\mu \mathbf{z}$. This mapping is interesting for several reasons, one of which is that an apparently classical object $\mathbf{n}$ allows an equivalent description in terms of the spinor $\mathbf{z}$, which can be viewed as the coherent-state wave function for spin-1/2. There are further advantages in looking at the spin dynamics from a CP¹ perspective, and it will be useful to have the same formulation when the DM interaction is added to the Hamiltonian.

In the angle representation, the DM interaction looks like

$$\begin{aligned}\mathbf{n}\cdot(\nabla\times\mathbf{n}) &= \sin\theta\cos\theta(\cos\phi\partial_x\phi+\sin\phi\partial_y\phi)\\ &\quad+(\sin\phi\partial_x\theta-\cos\phi\partial_y\theta)-\sin^2\theta\partial_z\phi.\end{aligned} \tag{3.76}$$

A matching CP¹ expression can be found by trial-and-error:

$$\mathbf{n}\cdot(\nabla\times\mathbf{n}) = -2\mathbf{n}\cdot\mathbf{a} - i\mathbf{z}^\dagger(\boldsymbol{\sigma}\cdot\nabla)\mathbf{z} + i(\nabla\mathbf{z}^\dagger)\cdot\boldsymbol{\sigma}\mathbf{z}. \tag{3.77}$$

Although the individual terms on the right-hand side are not gauge-invariant under $\mathbf{z}\to e^{if}\mathbf{z}$, the sum (3.77) is. When the exchange, DM, and Zeeman interactions are all combined, we arrive at the CP¹ representation of the HDMZ Hamiltonian

$$\begin{aligned}H &= \frac{J}{2}\sum_{\mu}(\partial_\mu\mathbf{n})\cdot(\partial_\mu\mathbf{n}) + D\mathbf{n}\cdot\nabla\times\mathbf{n} - \mathbf{B}\cdot\mathbf{n}\\ &= 2J\sum_{\mu}\left(\partial_\mu\mathbf{z}^\dagger + ia_\mu\mathbf{z}^\dagger - i\kappa\mathbf{z}^\dagger\sigma_\mu\right)\left(\partial_\mu\mathbf{z} - ia_\mu\mathbf{z} + i\kappa\sigma_\mu\mathbf{z}\right) - \mathbf{B}\cdot\mathbf{z}^\dagger\boldsymbol{\sigma}\mathbf{z}\\ &= 2J\sum_{\mu}(D_\mu\mathbf{z})^\dagger(D_\mu\mathbf{z}) - \mathbf{B}\cdot\mathbf{z}^\dagger\boldsymbol{\sigma}\mathbf{z}.\end{aligned} \tag{3.78}$$

In our discussions of the CP¹ formulation, κ will be defined temporarily as $\kappa\equiv D/2J$, half the value of the earlier definition for the spiral wave vector $k=\kappa$. The covariant derivative,

$$D_\mu = \partial_\mu - ia_\mu + i\kappa\sigma_\mu \tag{3.79}$$

now contains a nonabelian gauge field proportional to the Pauli matrix σ_μ due to the DM interaction. Although the nonabelian gauge field part is nondynamic (actually a constant), there is still a nontrivial nonabelian flux associated with it. Invoking the nonabelian gauge theory formula, the field tensor obtained from the gauge potential is

$$F_{\mu\nu} = i[D_\mu, D_\nu] = f_{\mu\nu} + 2\kappa^2 \varepsilon_{\mu\nu\lambda}\sigma_\lambda, \tag{3.80}$$

where the abelian part of the flux $f_{\mu\nu}$ is the usual one, $f_{\mu\nu} = \partial_\mu a_\nu - \partial_\nu a_\mu$.

In Sect. 3.4, we saw that the skyrmion lattice could be interpreted as a multi-spiral state. We will soon show how the same skyrmion crystal state can be derived in the CP^1 formulation. Before doing so, let us try to build up a little intuition for thinking about various spin structures in the CP^1 language. The simplest spin structure is that of a ferromagnet $\mathbf{n}_0 = (0, 0, 1)$, whose CP^1 representation is $\mathbf{z}_0 = (1, 0)$ such that $\mathbf{n}_0 = \mathbf{z}_0^\dagger \boldsymbol{\sigma} \mathbf{z}_0$. Less trivial is the spiral spin state with the wavevector $\mathbf{k} = k\hat{k}$, which has the CP^1 expression

$$\begin{aligned} \mathbf{z} &= e^{i(\boldsymbol{\sigma}\cdot\hat{k})(\mathbf{k}\cdot\mathbf{r})/2}\mathbf{z}_0, \\ \mathbf{n} &= \mathbf{z}^\dagger \boldsymbol{\sigma} \mathbf{z} = \mathbf{n}_0 \cos(\mathbf{k}\cdot\mathbf{r}) + (\mathbf{n}_0 \times \hat{k}) \sin(\mathbf{k}\cdot\mathbf{r}). \end{aligned} \tag{3.81}$$

The only requirement in deducing the second line is that the initial spin vector $\mathbf{n}_0 = \mathbf{z}_0 \boldsymbol{\sigma} \mathbf{z}_0$ needs to be orthogonal to $\mathbf{k}$.

Now, let us try to write down the single skyrmion state in the CP^1 language. We begin with a general observation that an arbitrary vector $\mathbf{n}$ has the CP^1 equivalent

$$\mathbf{z} = e^{i(\theta/2)(\boldsymbol{\sigma}\cdot\hat{\phi})}\mathbf{z}_0 = \begin{pmatrix} \cos[\theta/2] \\ e^{i\phi}\sin[\theta/2] \end{pmatrix}, \tag{3.82}$$

where $\hat{\phi} = (-\sin\phi, \cos\phi, 0)$. A single skyrmion spin configuration can then be obtained by choosing, for instance, the angle ϕ of the spin to coincide with the azimuthal angle $\varphi = \arctan(y/x)$ in the plane, and $\theta(r)$ to be a smooth function of the radial coordinate r with $\theta(0) = \pi$, and $\theta(\infty) = 0$. The vector potential for the single skyrmion configuration is then

$$\mathbf{a} = -i\mathbf{z}^\dagger \nabla \mathbf{z} = \frac{\hat{\varphi}}{2r}\Big(1 - \cos[\theta(r)]\Big), \tag{3.83}$$

where $\hat{\varphi} = (-\sin\varphi, \cos\varphi, 0)$. In the CP^1 language, the abelian part of the flux originating from this vector potential is

$$\nabla_2 \times \mathbf{a} = \frac{1}{2r}\sin[\theta(r)]\theta'(r). \tag{3.84}$$

which agrees with the flux obtained from the alternative definition given in (2.41).

The challenge now is to figure out whether the skyrmion crystal state allows a CP^1 formulation. Specifically, we are interested to know if such state will emerge as the saddle-point solution of the CP^1 chiral Hamiltonian (3.78). The saddle-point equation is found by supplementing the HDMZ Hamiltonian with the Lagrange multiplier $\lambda(\mathbf{z}^\dagger \mathbf{z} - 1)$ to impose the constraint $\mathbf{z}^\dagger \mathbf{z} = 1$. Setting the functional derivative of the modified Hamiltonian with respect to $\mathbf{z}^\dagger$ to be zero yields the saddle-point equation

$$2J(\nabla - i\mathbf{a} + i\kappa\boldsymbol{\sigma})^2\mathbf{z} + 4iJ\kappa(\mathbf{n}\cdot\nabla)\mathbf{z} + (\mathbf{B}\cdot\boldsymbol{\sigma})\mathbf{z} = \lambda\mathbf{z}. \tag{3.85}$$

This equation is extremely complex and heavily nonlinear. Nonetheless, one can check that the three example configurations considered above: ferromagnetic, spiral, and single skyrmion, all satisfy the saddle-point equation with an appropriate choice of λ.

Some simplifications can be achieved when B is assumed to be slightly less than B_{c2}, the upper critical field at which the skyrmion crystal phase gives way to a fully ferromagnetic state. There we can assume that the dominant part of $\mathbf{z}$ describes the uniform component, $\mathbf{z}_{\mathrm{FM}} = (\sqrt{m}\ \ 0)^T$, and the residual part, $\mathbf{z} - \mathbf{z}_{\mathrm{FM}}$, is used to describe the skyrmion lattice. In that case, we can take $\mathbf{n} \approx m\hat{z}$ to a first approximation, and hence that $\mathbf{n}\cdot\nabla\mathbf{z} \approx m\hat{z}\cdot\nabla\mathbf{z} = 0$, since the spatial gradient in the case of the two-dimensional lattice is in the xy-plane. Thanks to this simplification, (3.85) becomes

$$2J\left(\nabla - i\mathbf{a} + i\kappa\boldsymbol{\sigma}\right)^2\mathbf{z} + (\mathbf{B}\cdot\boldsymbol{\sigma})\mathbf{z} = \lambda\mathbf{z}. \tag{3.86}$$

The appearance of the U(1) gauge potential $\mathbf{a}$ in the CP^1 theory reminds us of a similar problem in the GL theory of a superconductor, subject to the constant magnetic field $\mathbf{B} = \nabla \times \mathbf{A}$. In that problem, ignoring the fourth-order term of the condensate order parameter $|\psi|^4$ when the system is sufficiently close to the upper critical field, the linearized saddle-point equation following from the GL theory is

$$-\frac{1}{2}(\nabla - i\mathbf{A})^2\psi = \lambda\psi, \tag{3.87}$$

which is exactly like the problem of electrons subject to a constant magnetic field. Solutions to this equation form Landau levels, in which the lowest-energy level is spanned by states of the form $\psi(x, y) = f(z)$ multiplied by a Gaussian envelope. The polynomial function $f(z)$ can be an arbitrary function of the complex variable $z = x + iy$. In this language, then, a single quantized vortex is described by the wave-function $f(z) = z$. Abrikosov proposed a particular kind of holomorphic function that satisfied the periodic properties [35],

$$f(z + a_1) = f(z) = f(z + a_2), \tag{3.88}$$

where a_1 and a_2 are the two complex numbers describing the vortex lattice vectors. The Abrikosov solution yields a periodic array of zeroes $f(z) = 0$ corresponding to the centers of the vortices, forming what is famously known as the Abrikosov vortex lattice.[7]

Abrikosov solved the problem in the Landau gauge $\mathbf{A} = Bx\hat{y}$ instead, and found the solution

$$\psi(x, y) = \sum_{n=-\infty}^{\infty} c_n e^{2\pi i n y / l_y} e^{-(x-nl_x)^2/2\xi^2}, \tag{3.89}$$

for some constants c_n. Here, ξ is the correlation length of a superconductor, while l_x and l_y are the inter-vortex separations in the x- and y-directions, respectively, and satisfy $l_x l_y = h/|2q_e|B = l_B^2$, where l_B is the magnetic length.

Knowing that the vector potential $\mathbf{a}$ of a single skyrmion yields an emergent field proportional to $\nabla \times \mathbf{a}$ (much like the diamagnetic field generated by the superconducting vortex), and that an array of skyrmions generates fields that, on average, look uniform, it is tempting to look for a solution of the CP^1 saddle-point equation (3.86) that resembles the Abrikosov solution. Note that there are some complications in the CP^1 problem for skyrmions that were absent in Abrikosov's treatment of vortices, i.e., the Zeeman field $(\mathbf{B} \cdot \boldsymbol{\sigma})\mathbf{z}$ and the nonabelian piece $\kappa\boldsymbol{\sigma}$. Putting these differences aside, we find that most of the tricks that Abrikosov invented to solve his problem also works well for the CP^1 skyrmion problem.

The emergent magnetic field will be denoted $\mathbf{b} = \nabla \times \mathbf{a}$ to distinguish it from the external field $\mathbf{B}$. With $\mathbf{B} = B\hat{z}$, the emergent field direction is assumed to be its opposite, $\mathbf{b} = -b\hat{z}$, in anticipation of the same kind of diamagnetic response that takes place in type-II superconductors. The spatial variation of $\mathbf{b}$ over the distance of one skyrmion is ignored in favor of a uniform field, so that b becomes a constant. Choosing the Landau gauge, the emergent vector potential $\mathbf{a}$ is given by $\mathbf{a} = (0, -bx, 0)$, and the translational symmetry of the CP^1 solution along the y-direction implies

$$\mathbf{z}(x, y) = e^{iky}\mathbf{z}(x). \tag{3.90}$$

The x-dependent part $\mathbf{z}(x)$ then obeys the equation

$$2J\left(\partial_x + i\kappa\sigma_x\right)^2 \mathbf{z} - 2J\left(k + bx - \kappa\sigma_y\right)^2 \mathbf{z} + B\sigma_z \mathbf{z} = \lambda \mathbf{z}. \tag{3.91}$$

From dimensional analysis it is clear that b gives rise to the effective magnetic length

$$l_b = 1/\sqrt{b}. \tag{3.92}$$

It is enlightening to rewrite (3.91) in terms of the following creation and annihilation operators:

[7] For readers wishing to follow up on Abrikosov's derivation, I recommend Tinkham's textbook on superconductivity [35].

$$a = -\frac{i}{\sqrt{2}}\left(l_b\frac{\partial}{\partial x} + \frac{x_k}{l_b}\right),$$
$$a^\dagger = \frac{i}{\sqrt{2}}\left(-l_b\frac{\partial}{\partial x} + \frac{x_k}{l_b}\right), \tag{3.93}$$

where $x_k = x + kl_b^2$. With these operators, the pair of differential equations (3.91) may be written as

$$b\left(a^\dagger a + \frac{1}{2}\right)z_1 - \sqrt{2b}\kappa a z_2 = \frac{-\lambda + B}{2J}z_1,$$
$$b\left(a^\dagger a + \frac{1}{2}\right)z_2 - \sqrt{2b}\kappa a^\dagger z_1 = -\frac{\lambda + B}{2J}z_2, \tag{3.94}$$

which are in terms of the two components of $\mathbf{z}(x) = (z_1(x)\, z_2(x))$. From the structure of these equations, it is easy to guess the solution,

$$\mathbf{z}_k(x) = \begin{pmatrix} \phi_n(x_k/l_b) \\ id_n\phi_{n+1}(x_k/l_b) \end{pmatrix},$$

which involves the normalized n-th and $(n+1)$-th harmonic oscillator wavefunctions. The coefficient d_n is determined from the equations

$$b\left(n + \frac{1}{2}\right) - \kappa\sqrt{2b(n+1)}\,d_n = \frac{-\lambda_n + B}{2J},$$
$$b\left(n + 1 + \frac{1}{2}\right)d_n - \kappa\sqrt{2b(n+1)} = -\frac{\lambda_n + B}{2J}d_n. \tag{3.95}$$

The Lagrange multiplier λ_n depends on the given Landau level index n. Solving the coupled linear equations yields

$$d_n = \frac{2\kappa\sqrt{2b(n+1)}}{b + B/2J + \sqrt{(b + B/2J)^2 + 8\kappa^2 b(n+1)}}, \tag{3.96}$$

and the solution of the saddle-point problem is complete.

The most likely ground state is achieved for $n = 0$, and we shall now drop the Landau level index assuming that this state is realized. We can then write

$$\mathbf{z}(x_k, y) = e^{iky}\begin{pmatrix} \phi_0(x_k/l_b) \\ d_0\,\phi_1(x_k/l_b) \end{pmatrix} = e^{iky}e^{-x_k^2/2l_b^2}\begin{pmatrix} 1 \\ \sqrt{2}id_0x_k/l_b \end{pmatrix}. \tag{3.97}$$

As in the Abrikosov problem, solutions with different values of k are degenerate and can be grouped into a linear combination

$$\mathbf{z}(x, y) = \sum_k c_k z(x_k, y) = \sum_k c_k e^{iky} e^{-x_k^2/2l_b^2} \begin{pmatrix} 1 \\ \sqrt{2} i d_0 x_k / l_b \end{pmatrix}. \quad (3.98)$$

Let us see how this compares with Abrikosov's solution

$$\psi(x, y) = \sum_{n=-\infty}^{\infty} c_n e^{2\pi i n y / l_y} e^{-(x - n l_x)^2 / 2\xi^2}. \quad (3.99)$$

The upper component of the CP[1] solution (3.98) is precisely the Abrikosov's vortex solution, provided we push the analogy further by taking $k = 2\pi n / l_y$, n=integer. Then it automatically follows that

$$\begin{aligned} k l_b^2 &= \frac{2\pi n}{l_y b} = n l_x, \\ l_x l_y &= \frac{2\pi}{b} = 2\pi l_b^2, \end{aligned} \quad (3.100)$$

and our solution becomes

$$\mathbf{z}(x, y) = \sum_{n=-\infty}^{\infty} c_n e^{2\pi i n y / l_y} e^{-x_n^2/2l_b^2} \begin{pmatrix} 1 \\ \sqrt{2} i d_0 x_n / l_b \end{pmatrix}, \quad (3.101)$$

where $x_n = x + n l_x$. One can even borrow the values of c_n from Abrikosov's work

$$\begin{aligned} c_n &= 1 \ (n = \text{even}) \\ &= i \ (n = \text{odd}), \end{aligned} \quad (3.102)$$

and use $l_y = \sqrt{3} l_x / 2$ to construct the triangular lattice.

How many different n's are allowed in a sample of finite width L_x? Calling that number N, we have $N l_x = L_x$ or $N = L_x / l_x$. Furthermore, we can consider what happens if the normalization condition $\mathbf{z}^\dagger \mathbf{z} = 1$ is replaced by the more relaxed, average condition

$$\frac{1}{L_x L_y} \int dx dy \, \mathbf{z}^\dagger \mathbf{z} = 1 = \frac{\sqrt{\pi}}{2 L_x} l_b (1 + d_0^2) \sum_n |c_n|^2. \quad (3.103)$$

Assuming that the $|c_n|^2$'s are the same for all the n's yields

$$|c_n| = \left(\frac{2 l_x}{(1 + d_0^2) l_b \sqrt{\pi}} \right)^{1/2}. \quad (3.104)$$

We can then drop the normalization and rewrite our final form of the topological lattice solution as

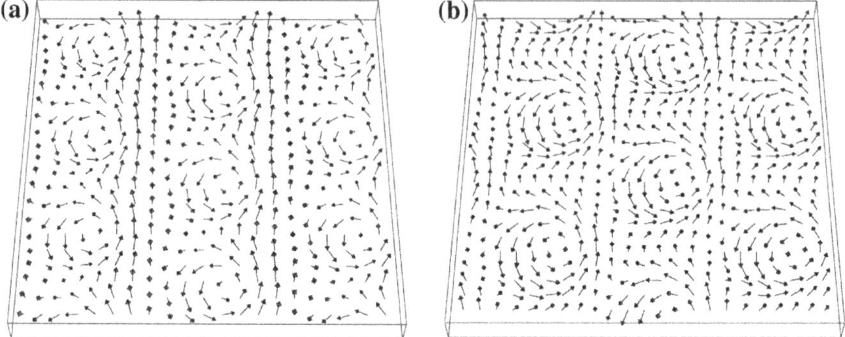

Fig. 3.9 Skyrmion crystal configuration obtained from **a** variational CP[1] calculation and **b** the Monte Carlo calculation. Reproduced with permission from Ref. [36]

$$\mathbf{z}(x,y)=\sqrt{\frac{2l_x}{(1+d_0^2)l_b\sqrt{\pi}}}\sum_{n=-\infty}^{\infty}c_n e^{2\pi i n y/l_y}e^{-x_n^2/2l_b^2}\begin{pmatrix}1\\ \sqrt{2}id_0x_n/l_b\end{pmatrix}, \quad (3.105)$$

where the c_n's are unit modulus constants given in (3.102).

Inspection of the CP^1 solution (3.105) and the definition of d_0 (3.96) reveals that there is one remaining variational parameter b in the CP^1 wave function. The emergent field strength b can be fixed by feeding the $\mathbf{z}(x,y)$ in (3.105) back into the CP^1 Hamiltonian (3.78), to obtain the variational energy of the skyrmion crystal

$$E=2J\left(2\kappa^2-\frac{4\sqrt{2b}\kappa d_0}{1+d_0^2}+\frac{1+3d_0^2}{1+d_0^2}b\right)-B\frac{1-d_0^2}{1+d_0^2}, \quad (3.106)$$

and taking $\partial E/\partial b=0$. With the value of b thus determined, one can plot the resulting magnetization profile $\mathbf{n}=\mathbf{z}^\dagger\boldsymbol{\sigma}\mathbf{z}$ and compare it against the akyrmion lattice found from MC simulation. As shown in Fig. 3.9, the comparison is rather satisfactory.

It is nice to see the existence of a tight parallel between the quantum-mechanical wavefunction of the vortex lattice, and the quasi-wavefunction description of the classical skyrmion lattice solution through the CP^1 mapping. This is not to say that the physics of the two are the same. The classical magnet and the quantum condensate are two wildly different physical systems. Nevertheless, in the following chapter, we will show that another close parallel exists between the two systems in the equations of motion obeyed by their respective topological objects, i.e., a vortex in a superconductor or superfluid, and a skyrmion in a ferromagnet.

References

1. Volovik, G.E.: Exotic Properties of Superfluid ^{3}He. World Scientific (1992)
2. Volovik, G.E.: The Universe in a Helium Droplet. Oxford University Press, New York (2009)
3. Sarma, S.D., Pinczuk, A.: Perspective in Quantum Hall Effects: Novel Quantum Liquids in Low-dimensional Semiconductor Structures. Wiley, VCH (1996)

4. Malozemoff, A.P., Slonczewski, J.C.: Magnetic Domain Walls in Bubble Materials. Academic Press, New York (1979)
5. Eschenfelder, A.H.: Magnetic Bubble Technology. Springer Series in Solid-State Sciences. Springer, Berlin (1981)
6. Ishikawa, Y., Tajima, K., Bloch, D., Roth, M.: Helical spin structure in manganese silicide MnSi. Solid State Commun. **86–88B**, 525 (1976)
7. Ishikawa, Y., Shirane, G., Tarvin, J.A., Khogi, M.: Magnetic excitations in the weak itinerant ferromagnet MnSi. Phys. Rev. B **16**, 4956 (1977)
8. Bak, P., Jensen, M.H.: Theory of helical magnetic structures and phase transitions in MnSi and FeGe. J. Phys. C: Solid State Phys. **13**, L881 (1980)
9. Dzyaloshinsky, I.E.: A thermodynamic theory of weak ferromagnetism of antiferromagnetics. J. Phys. Chem. Solids **4**, 241 (1958)
10. Dzyaloshinskii, I.E.: Theory of helicoidal structures in antiferromagnets. I. Nonmetals. Sov. Phys. JETP **19**, 960 (1964)
11. Dzyaloshinskii, I.E.: The theory of helicoidal structures in antiferromagnets. II. metals. Sov. Phys. JETP **20**, 223 (1965)
12. Moriya, T.: New mechanism of anisotropic superexchange interaction. Phys. Rev. Lett. **4**, 228 (1960)
13. Moriya, T.: Anisotropic superexchange interaction and weak ferromagnetism. Phys. Rev. **120**, 91 (1960)
14. Bogdanov, A.N., Yablonskii, D.A.: Thermodynamically stable vortices in magnetically ordered crystals. The mixed state of magnets. Sov. Phys. JETP **68**, 101 (1989)
15. Bogdanov, A.N., Hubert, A.: Thermodynamically stable magnetic vortex states in magnetic crystals. J. Mag. Mag. Mat. **138**, 255 (1994)
16. Bogdanov, A.N., Hubert, A.: The stability of vortex-like structures in uniaxial ferromagnets. J. Mag. Mag. Mat. **195**, 182 (1999)
17. Bogdanov, A.N., Rößler, U.K.: Chiral symmetry breaking in magnetic thin films and multilayers. Phys. Rev. Lett. **87**, 037203 (2001)
18. Bogdanov, A.N., Rößler, U.K., Wolf, M., Müuller, K.H.: Magnetic structures and reorientation transitions in noncentrosymmetric uniaxial antiferromagnets. Phys. Rev. B **66**, 214410 (2002)
19. Rößler, U.K., Bogdanov, A.N., Pfleiderer, C.: Spontanoues skyrmion ground states in magnetic metals. Nature **442**, 797 (2006)
20. Rößler, U.K., Leonov, A.A., Bogdanov, A.N.: Chiral skyrmionic matter in non-centrosymmetric magnets. J. Phys: Conf. Ser. **303**, 012105 (2011)
21. Pfleiderer, C., Reznik, D., Pintschovius, L., Löhneysen, H.v., Garst, M., Rosch, A.: Partial order in the non-Fermi-liquid phase of MnSi. Nature **427**, 227 (2004)
22. Binz, B., Vishwanath, A., Aji, V.: Theory of the helical spin crystal: a candidate for the partially ordered state of MnSi. Phys. Rev. Lett. **96**, 207202 (2006)
23. Binz, B., Vishwanath, A.: Theory of helical spin crystals: Phases, textures, and properties. Phys. Rev. B **74**, 214408 (2006)
24. Mühlbauer, S., Binz, B., Jonietz, F., Pfleiderer, C., Rosch, A., Neubauer, A., Georgii, R., Böni, P.: Skyrmion lattice in a chiral magnet. Science **323**, 915 (2009)
25. Lee, M., Onose, Y., Tokura, Y., Ong. N.P.: Hidden constant in the anomalous Hall effect of high-purity magnet MnSi. Phys. Rev. B **75**, 172403 (2007)
26. Yi, S.D., Onoda, S., Nagaosa, N., Han, J.H.: Skyrmions and anomalous Hall effect in a Dzyloshinskii-Moriya spiral magnet. Phys. Rev. B **80**, 054416 (2009)
27. Yu, X.Z., Onose, Y., Kanazawa, N., Park, J.H., Han, J.H., Matsui, Y., Nagaosa, N., Tokura, Y.: Real-space observation of a two-dimensional skyrmion crystal. Nature **465**, 901 (2010)
28. Jeong, T., Pickett, W.E.: Implications of the B20 crystal structure for the magnetoelectronic structure of MnSi. Phys. Rev. B **70**, 075114 (2004)
29. Nagaosa, N., Tokura, Y.: Topological properties and dynamics of magnetic skyrmions. Nat. Nanotech. **8**, 899 (2013)
30. Rohart, S., Thiaville, A.: Skyrmion confinement in ultrathin film nanostructures in the presence of Dzyaloshinskii-Moriya interaction. Phys. Rev. B **88**, 184422 (2013)

31. Han, J.H., Park, J.H.: Zero-temperature phases for chiral magnets in three dimensions. Phys. Rev. B **83**, 184406 (2011)
32. Buhrandt, S., Fritz, L.: Skyrmion lattice phase in three-dimensional chiral magnets from Monte Carlo simulations. Phys. Rev. B **88**, 195137 (2013)
33. Yang, S.G., Liu, Y.H., Han, J.H.: Formation of a topological monopole lattice and its dynamics in three-dimensional chiral magnets. Phys. Rev. B **94**, 054420 (2016)
34. Kanazawa, N., Onose, Y., Arima, T., Okuyama, D., Ohoyama, K., Wakimoto, S., Kakurai, K., Ishiwata, S., Tokura, Y.: Large topological hall effect in a short-period Helimagnet MnGe. Phys. Rev. Lett. **106**, 156603 (2011)
35. Tinkham, M.: Introduction to Superconductivity, 2nd edn. Dover books on physics. Dover Publications, New York (2004)
36. Han, J.H., Zang, J., Yang, Z., Park, J.H., Nagaosa, N.: Skyrmion lattice in two-dimensional chiral magnet. Phys. Rev. B **82**, 094429 (2010)

Chapter 4
Skyrmion Equation of Motion

In the previous chapter, the energetic stability of a single skyrmion in chiral magnets was examined using the Ginzburg-Landau theory, allowing us to understand the thermodynamic stability of the skyrmion lattice from a number of different perspectives. In this chapter, we move on to the dynamics of skyrmions. There are many parallels between the dynamical theories of vortices in superfluids and skyrmions in chiral magnets, both of which bear a resemblance to the guiding center dynamics of charged particles in magnetic fields. All three theories can be phrased in the language of Berry phase dynamics. First, we derive the equation of motion governing the dynamics of a single vortex from the Berry phase perspective and by using the theory of duality. The dynamics of skyrmions are developed next, starting from the familiar Berry phase spin-Lagrangian. It turns out that the single-particle dynamics of these two topological objects are remarkably similar, if not identical. The fundamental equation of motion for the skyrmion is applied to a number of situations, including the gyration mode in a nanodisk and the collective mode of the skyrmion lattice.

4.1 Effective Equation of Motion of a Quantized Vortex

4.1.1 Gross–Pitaevskii Action Approach

Quantized vortices in superfluids and superconductors are probably the first common instances of "topological particles" encountered in condensed matter physics.[1] The original idea for the quantized vortex in superfluids is widely credited to Lars Onsager,

[1]Much earlier, Lord Kelvin proposed in the late nineteenth century that atoms could be none other than topological objects known as knots [1]. His proposal subsequently prompted a wealth of

J.H. Han, *Skyrmions in Condensed Matter*, Springer Tracts in Modern Physics 278, https://doi.org/10.1007/978-3-319-69246-3_4

who made the remark in 1949 [2].[2] Vortices are also familiar and important objects in classical hydrodynamics. A whole chapter of Lamb's classic book on hydrodynamics is devoted to it [4]. Vortices take on a more "particle-like" character in superfluids due to the quantized integer number attached to them. Just like the charge of an elementary particle, the quantized vorticity endows such topological "particles" with a certain robustness.

A quantized vortex is a particular example of an emergent particle (skyrmions being another). It represents a collective state of the underlying particles, and yet, in many respects, is amenable to a single-particle description with its own equation of motion. An important step forward in the theory of vortex dynamics came from the realization that the Magnus force, i.e., the force acting transverse to the vortex's velocity against the medium, can be interpreted as a manifestation of the Berry phase. As an emergent particle, many of its properties can, in principle, be extracted from the underlying theory of its constituent microscopic particles. For quantized vortices, that underlying theory is the Gross–Pitaevskii (GP) equation,

$$i\hbar\frac{\partial\psi(\mathbf{r},t)}{\partial t}=\left(-\frac{\hbar^2}{2m}\nabla^2+U(\mathbf{r})+g|\psi(\mathbf{r},t)|^2\right)\psi(\mathbf{r},t), \tag{4.1}$$

or the equivalent GP Lagrangian:

$$L=i\hbar\psi^*\partial_t\psi-\frac{\hbar^2}{2m}(\nabla\psi^*\cdot\nabla\psi)-U(\mathbf{r})|\psi|^2-\frac{g}{2}(|\psi|^2-\rho_0)^2. \tag{4.2}$$

The complex-valued function ψ is known as the condensate wavefunction. The interaction term of strength g describes the cost of density fluctuations around the mean value ρ_0.

One can rewrite the Lagrangian in the so-called hydrodynamic form. First, the condensate wavefunction ψ is written in terms of the hydrodynamic variables, ρ (density of superfluid) and the U(1) phase $e^{i\theta}$, as $\psi=\sqrt{\rho}e^{i\theta}$. Substituting the hydrodynamic fields back into the GP Lagrangian yields the same Lagrangian in hydrodynamic form:

$$L=-\hbar\rho\dot{\theta}-\frac{\hbar^2}{2m}\left(\frac{1}{4}\frac{[\nabla\rho]^2}{\rho}+\rho[\nabla\theta]^2\right)-U(\mathbf{r})\rho-\frac{g}{2}(\rho-\rho_0)^2. \tag{4.3}$$

---

theoretical and mathematical ideas, including Skyrme's own idea to treat elementary particles as topological objects.

[2]Readers interested in accounts of the early history of quantized vortices in superfluids are encouraged to seek out D. Thouless' book [3]. The book provides a succinct review of each of the main topological themes in condensed matter physics, along with an account of the significant contributions he made to some of those fields that earned him the 2016 Nobel Prize in physics together with Michael Kosterlitz and Duncan Haldane.

There are many uses of the hydrodynamic Lagrangian, but here we focus on one issue: is it possible to deduce the Lagrangian of a single vortex by starting from the hydrodynamic Lagrangian?

A typical single-particle Lagrangian should include the kinetic energy that is quadratic in the velocity and linear in the mass. Upon inspection of the hydrodynamic action (4.3), however, one discovers that there is no quadratic time derivative term. As we will show, the second-order term in $\dot{\theta}$ can be "generated" by the coupling of the phase derivative $\dot{\theta}$ to the density fluctuation $\delta\rho$ through the first term in the action (4.3). The condensate density is constrained around the mean value ρ_0 by the interaction term, allowing one to write $\rho = \rho_0 + \delta\rho$ and treat $\delta\rho$ as the dynamical variable. In the homogeneous background with $U(\mathbf{r}) = 0$, keeping the lowest order terms in $\delta\rho$ gives the hydrodynamic action

$$\begin{aligned} L &= -\hbar\dot{\theta}\delta\rho - \frac{\hbar^2}{2m}\left(\frac{1}{4}\frac{[\nabla\delta\rho]^2}{\rho_0} + \rho_0[\nabla\theta]^2\right) - \frac{g}{2}(\delta\rho)^2 \\ &= -\hbar\dot{\theta}\delta\rho - \frac{1}{2}\delta\rho\left(g - \frac{\hbar^2}{4m\rho_0}\nabla^2\right)\delta\rho - \frac{\hbar^2\rho_0}{2m}(\nabla\theta)^2. \end{aligned} \tag{4.4}$$

The key observation here is that both dynamical fields θ and $\delta\rho$ appear to quadratic order, at most, and we know that quadratic field theory can be solved (in principle at least) exactly. Let us symbolically define the inverse Green's function, $G^{-1} = g - (\hbar^2/4m\rho_0)\nabla^2$, and complete the square

$$\begin{aligned} &-\hbar\dot{\theta}\delta\rho - \frac{1}{2}\delta\rho G^{-1}\delta\rho \\ &= -\frac{1}{2}[\delta\rho + \hbar\dot{\theta}G]G^{-1}[\delta\rho + \hbar G\dot{\theta}] + \frac{\hbar^2}{2}\dot{\theta}G\dot{\theta}. \end{aligned} \tag{4.5}$$

In the field-theoretical parlance, the density fluctuation part can be "integrated out" by performing the Gaussian integration, leaving behind an effective Lagrangian for θ alone:

$$L_{\text{eff}} = \frac{\hbar^2}{2}\int d^2\mathbf{r}d^2\mathbf{r}'\,\dot{\theta}(\mathbf{r})G(\mathbf{r}-\mathbf{r}')\dot{\theta}(\mathbf{r}') - \frac{\hbar^2\rho_0}{2m}\int d^2\mathbf{r}\,(\nabla\theta)^2. \tag{4.6}$$

Since we are mainly interested in the vortex dynamics in two dimensions, we have made the integral explicitly two-dimensional.

The effective Lagrangian derived in (4.6) is nonlocal in space, as often happens once the "integrating-out" procedure has been performed. Under the assumption that the Green's function is sufficiently local in space, one can make the following approximation:

$$G(\mathbf{r}) \approx G(\mathbf{0}) = \int \frac{d^2\mathbf{k}}{(2\pi)^2} G(\mathbf{k}) = \int_{k_m}^{k_M} \frac{2\pi k dk}{(2\pi)^2} \frac{1}{g + (\hbar^2/4m\rho_0)k^2}$$
$$= \frac{m}{\pi\hbar^2} \ln\left[\frac{k_M^2 + 4m\rho_0 g/\hbar^2}{k_m^2 + 4m\rho_0 g/\hbar^2}\right]. \tag{4.7}$$

Two cut-offs k_m and k_M have been introduced to control the integral. We have the effective action for θ which is local:

$$L_{\text{eff}} = \frac{m^*}{2\pi}\dot{\theta}^2 - \frac{\hbar^2\rho_0}{2m}(\nabla\theta)^2, \quad m^* = m \ln\left[\frac{k_M^2 + 4m\rho_0 g/\hbar^2}{k_m^2 + 4m\rho_0 g/\hbar^2}\right]. \tag{4.8}$$

This expression is the well-known phason action for the superfluid. The linear dispersion of the phase field $\omega_{\mathbf{k}} \sim |\mathbf{k}|$ comes from the Goldstone mode tied to the breaking of the gauge symmetry in the ground state. We will not worry a great deal about justifying the procedure used in deriving the phason Lagrangian, and instead use it as our starting point for performing the next task.

Up to now, no reference has been made to the single vortex excitation. All we have shown is how the second-order time derivative in $\dot{\theta}$ follows from a certain integrating-out procedure. The effective action of a single vortex can be deduced from (4.8) by constructing a vortex ansatz for the condensate wavefunction (writing $A = \sqrt{\rho}$ for the amplitude)

$$\psi(\mathbf{r}, \mathbf{R}(t)) = A(\mathbf{r} - \mathbf{R}(t))e^{i\theta(\mathbf{r}-\mathbf{R}(t))},$$
$$\theta(\mathbf{r} - \mathbf{R}(t)) = \arctan\left(\frac{y - Y(t)}{x - X(t)}\right). \tag{4.9}$$

The crucial assumption here is that time dependence arises solely from that of the vortex coordinates $\mathbf{R}(t) = (X(t), Y(t))$. With this ansatz, the first-order time derivative of the phase angle becomes $\dot{\theta} = -\dot{\mathbf{R}} \cdot \nabla\theta$, and the kinetic energy of the phase field in terms of the vortex velocity $\dot{\mathbf{R}}$ becomes

$$\frac{m^*}{2\pi} \int d^2\mathbf{r}\, \dot{\theta}^2 \Rightarrow \frac{m^*}{2\pi} \dot{R}_i \dot{R}_j \int d^2\mathbf{r}\, \partial_i\theta\partial_j\theta$$
$$= \frac{1}{2}\frac{m^*}{2\pi}\left[\int d^2\mathbf{r}\, (\nabla\theta)^2\right]\dot{\mathbf{R}}^2. \tag{4.10}$$

The integral $\int d^2\mathbf{r}\, (\nabla\theta)^2 = 2\pi \ln(L/a)$ gives the well-known logarithmic divergence (bounded by the linear dimension of the container L and the atomic spacing a), which carries over to the effective mass of a single vortex,

$$M_v = m^* \ln(L/a). \tag{4.11}$$

Starting from the Gross–Pitaevskii Lagrangian, we were able to work our way to deducing the kinetic energy of the vortex $M_v\dot{\mathbf{R}}^2/2$, and providing an estimate of the effective mass. On the other hand, the approach failed to produce the desired Berry phase in the action, which is linear in the vortex velocity $\dot{\mathbf{R}}$, and ultimately leads to the transverse Magnus force. In the next section, we outline a different track of thoughts that directly aims to unearth the Berry phase dynamics of the vortex.

4.1.2 Variational Wavefunction Approach

The single-vortex state as a many body boson state in the superfluid can be described by a variational wavefunction[3]

$$\begin{aligned}\Psi_v(\{\mathbf{r}_i\},\mathbf{R}) &= \Psi_0(\{\mathbf{r}_i\})A\left(\{\mathbf{r}_i-\mathbf{R}\}\right)\\ A(\{\mathbf{r}_i-\mathbf{R}\}) &= \prod_i\Big([(x_i-X)+i(y_i-Y)]f(|\mathbf{r}_i-\mathbf{R}|)\Big)f(|\mathbf{r}_i-\mathbf{R}|).\end{aligned} \quad (4.12)$$

The wavefunction $\Psi_0(\{\mathbf{r}_i\})$ describes the ground state of the superfluid. The vortex nature is encoded in the complex-valued prefactor $A(\{\mathbf{r}_i-\mathbf{R}\})$ of the wavefunction, which accounts for the phase winding of 2π of each boson particle at $\mathbf{r}_i=(x_i,y_i)$ around the vortex core at $\mathbf{R}=(X,Y)$ (corresponding to an angular momentum of $\hbar$ per particle), as well as through the radially dependent function $f(|\mathbf{r}_i-\mathbf{R}|)$ that vanishes as $\mathbf{r}_i\to\mathbf{R}$ to account for the density depletion at the core. A multi-vortex generalization, allowing for both positive and negative vorticities, is possible by multiplying the prefactor

$$A_\alpha(\{\mathbf{r}_i-\mathbf{R}_\alpha\}) = \prod_i\Big([(x_i-X_\alpha)+iq_\alpha(y_i-Y_\alpha)]f(|\mathbf{r}_i-\mathbf{R}_\alpha|)\Big) \quad (4.13)$$

for each vortex of topological charge $q_\alpha=\pm1$ located at $\mathbf{R}_\alpha$. The overall wavefunction for the multi-vortex state is then

$$\Psi_v(\{\mathbf{r}_i\},\{\mathbf{R}_\alpha\}) = \Psi_0(\{\mathbf{r}_i\})\prod_\alpha A_\alpha(\{\mathbf{r}_i-\mathbf{R}_\alpha\}). \quad (4.14)$$

Variational wavefunctions of this kind work best if the density depletion due to each vortex, $f(|\mathbf{r}_i-\mathbf{R}_\alpha|)$, occurs over a distance that is much less than the inter-vortex spacing. In such circumstances, each vortex can be viewed as a point-like object interacting with other point-like objects. On the other hand, the phase winding arising from the presence of vortices has a long-ranged effect and is additive, i.e., a particle

[3]The first instance of writing such a variational wavefunction is credited to Richard Feynman. See Feynman's works [5, 6] for a pedagogical exposition of his views on superfluids.

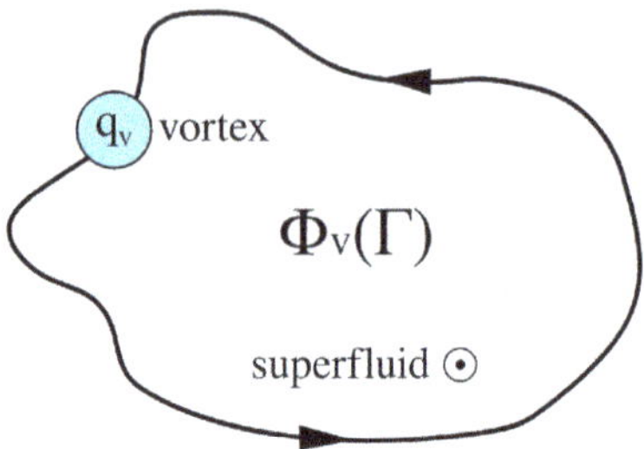

Fig. 4.1 Geometric phase of a vortex of charge q_v is calculated by considering the phase change in the variational vortex wavefunction accumulated while traversing a closed loop Γ in a superfluid background

encircling two vortices of charges q_1 and q_2 will experience the phase change of $2\pi(q_1+q_2)$ (Fig. 4.1).

In the previous section, we showed how to extract the Newtonian kinetic energy of a vortex from the microscopic GP theory. Now we show how to extract the other piece of dynamics tied to the Berry phase effect, from the variational wavefunction presented in (4.12).

As pointed out by Wu and Haldane [7] in 1986 and half a decade later by Ao and Thouless [8], we first raise the question: "What geometric phase is acquired by the wavefunction Ψ_v, as the vortex is dragged round a closed path?". In other words, one needs to calculate the geometric phase

$$\begin{aligned}\Phi_v(\Gamma) &= -i\oint_\Gamma d\mathbf{R}\cdot\langle\Psi_v(\mathbf{R})|\frac{\partial}{\partial\mathbf{R}}|\Psi_v(\mathbf{R})\rangle \\ &= -i\oint_\Gamma d\mathbf{R}\cdot\int d^2\mathbf{r}_1\cdots d^2\mathbf{r}_N\,\Psi_v^*(\{\mathbf{r}_i\},\mathbf{R})\frac{\partial}{\partial\mathbf{R}}\Psi_v(\{\mathbf{r}_i\},\mathbf{R}),\end{aligned} \tag{4.15}$$

as the vortex trajectory traces out a closed loop Γ in the superfluid background. It is typical in this kind of integrals that only the phase part of the wavefunction Ψ_v contributes, which comes from the vortex prefactor A.

Acting on the vortex prefactor $A(\{\mathbf{r}_i-\mathbf{R}\})$ in (4.12) gives

$$\frac{\partial}{\partial\mathbf{R}}A(\{\mathbf{r}_i-\mathbf{R}\}) = \left(\sum_i\frac{(1,iq_v)}{Z-z_i}\right)A(\{\mathbf{r}_i-\mathbf{R}\}), \tag{4.16}$$

where we also took account for the vortex charge q_v and introduced complex coordinates $Z=X+iq_vY$, $z_i=x_i+iq_vy_i$. In terms of the new coordinates, the Berry phase $\Phi_v(\Gamma)$ becomes the integral

$$\begin{aligned}\Phi_v(\Gamma) &= -i\int d^2\mathbf{r}_1\cdots d^2\mathbf{r}_N\left(\sum_i\oint_\Gamma\frac{dZ}{Z-z_i}\right)\rho_v(\{\mathbf{r}_i\}) \\ &= -i\int d^2\mathbf{r}\oint_\Gamma\frac{dZ}{Z-z}\rho_v(\mathbf{r}).\end{aligned} \tag{4.17}$$

In the second line, we have introduced the density matrix

$$\rho_v(\mathbf{r}) \equiv N \int d^2\mathbf{r}_2 \cdots d^2\mathbf{r}_N \; \rho_v(\{\mathbf{r}_i\}) \tag{4.18}$$

and made use of the symmetry of the wavefunction under the permutation of the particles' coordinates. The total number of particles in the condensate is assumed to be N. The contour integral $\oint_\Gamma$ leaves a residue $2\pi i q_v$ only if the particle resides in the interior of the loop $z \in S$ $(\partial S = \Gamma)$. Accordingly, the geometric phase acquired by the vortex is

$$\Phi_v(\Gamma) = 2\pi q_v \int_{\mathbf{r}\in S} d^2\mathbf{r} \; \rho_v(\mathbf{r}). \tag{4.19}$$

If the loop is much larger than the dimensions of a typical vortex, the density variation in the core region becomes a minor effect, and the above integral can be evaluated taking $\rho_v(\mathbf{r}) \approx \rho_0(\mathbf{r}) = \rho_0$, i.e., the density matrix for the uniform ground state. In that case, the Berry phase simply counts how many particles are enclosed by the loop:

$$\Phi_v(\Gamma) = 2\pi q_v N_\Gamma. \tag{4.20}$$

Drawing on the Aharonov-Bohm analogy, we see that the superfluid density effectively acts as a magnetic field on the vortex particle.

This conclusion, reached by considering the microscopic wavefunction, must be held up by any effective action description of the vortex dynamics. In other words, the single-vortex Lagrangian L_v must share a property that, on encircling a closed contour Γ, there is a phase equal to the one worked out in (4.20):

$$\frac{1}{\hbar} \oint_\Gamma L_v dt = 2\pi q_v N_\Gamma. \tag{4.21}$$

A Lagrangian that depends on the vortex position $\mathbf{R}$ with such property is easily guessed, because we know of an analogous situation for electrons subject to a magnetic field. The answer must be $L_v = (1/2)hq_v\rho_0\hat{z} \cdot (\mathbf{R} \times \dot{\mathbf{R}})$. In the vortex context, the equilibrium particle density ρ_0 replaces the magnetic field in accounting for the geometric phase. Together with the Newtonian kinetic energy $M_v\dot{\mathbf{R}}^2/2$ and some potential energy $V(\mathbf{R})$, we arrive at a complete "vortex-as-a-particle" Lagrangian

$$L_v = \frac{1}{2}M_v\dot{\mathbf{R}}^2 + \frac{1}{2}hq_v\rho_0\hat{z} \cdot (\mathbf{R} \times \dot{\mathbf{R}}) - V(\mathbf{R}). \tag{4.22}$$

The vortex mass estimation comes from the GP analysis of the previous section, while the geometric phase came from the variational wavefunction analysis. Through a compendium of analyses we have obtained an action that ultimately looks like that of an electron subject to the electric and magnetic fields.

The vortex action leads to the following equation of motion:

$$M_v \ddot{\mathbf{R}} = -\frac{\partial V}{\partial \mathbf{R}} + hq_v\rho_0 \dot{\mathbf{R}} \times \hat{z}, \tag{4.23}$$

which reads exactly like Lorentz's force equation for a charged particle in electrodynamics, despite the charge neutrality of the vortex. The popular jargon describing this kind of situation is "emergent electrodynamics". It is the topological charge of the vortex q_v that plays an analogous role as the electric charge. The fundamental reason for the similarity in their dynamics is that both phenomena are dictated by the Berry phase in their respective Lagrangians.

The vortex Hamiltonian is obtained from the standard Legendre transformation prescription. We define

$$\begin{aligned} P_X &= \partial L_v/\partial \dot{X} = M\dot{X} - hq_v\rho_0 Y/2, \\ P_Y &= \partial L_v/\partial \dot{Y} = M\dot{Y} + hq_v\rho_0 X/2, \end{aligned} \tag{4.24}$$

as the canonical momentum pair of the vortex, and consider the transform

$$\begin{aligned} H_v = \dot{X}P_X + \dot{Y}P_Y - L_v &= \frac{1}{2M}\left(P_X + \frac{1}{2}hq_v\rho_0 Y\right)^2 + \frac{1}{2M}\left(P_Y - \frac{1}{2}hq_v\rho_0 X\right)^2 \\ &= \frac{1}{2M}\left(\Pi_X^2 + \Pi_Y^2\right). \end{aligned} \tag{4.25}$$

This is the standard Hamiltonian for a charged particle in two dimensions subject to a uniform, perpendicular magnetic field. The pair of conjugate momenta (Π_X, Π_Y) satisfies the commutation relation

$$[\Pi_X, \Pi_Y] = 2\pi i\hbar^2 q_v\rho_0 \tag{4.26}$$

due to the canonical commutation rule $[X, P_X] = i\hbar = [Y, P_Y]$. A second pair of conjugate variables is provided by the guiding center coordinates (X_g, Y_g), which are related to the vortex coordinates (X, Y) by

$$\begin{aligned} X_g &= X + \frac{1}{hq\rho_0}\Pi_Y = \frac{1}{2}X + \frac{P_Y}{hq_v\rho_0}, \\ Y_g &= Y - \frac{1}{hq\rho_0}\Pi_X = \frac{1}{2}Y - \frac{P_X}{hq_v\rho_0}. \end{aligned} \tag{4.27}$$

Both guiding center coordinates commute with the Hamiltonian $[X_g, H] = [Y_g, H] = 0$ and obey the commutation

$$[X_g, Y_g] = -\frac{i}{2\pi q_v\rho_0}. \tag{4.28}$$

The commutation relation implies that the guiding center position has an uncertainty of the order of the inter-particle spacing in the superfluid, which is much smaller than the typical size of the vortex core. In practice, there is little need to sharply distinguish the two coordinates, $(X, Y) \approx (X_g, Y_g)$, since the "cyclotron radius" of the vortex is so small compared to the overall vortex size.

Just as a vortex resides in a superfluid background, skyrmions reside in a background of ferromagnetic spins. Although the physical media could not be more different, yet the effective dynamics of the skyrmion is entirely similar to that of the vortex. The analogy is so deep that one merely needs to replace the vortex charge q_v by the skyrmion charge Q_s and the superfluid density ρ_0 by the inverse lattice spacing squared, $\rho_0 \to 1/a^2$, as will be shown in the next section. Before doing that, we first try to complete the hydrodynamic theory of the vortex by introducing a duality theory, which captures the Berry phase aspect of the vortex dynamics in a natural way.

4.1.3 Duality Approach

The GP approach of Sect. 4.1.1. missed the important Berry phase contribution to the vortex action. The microscopic arguments by Wu and Haldane, and by Ao and Thouless gave convincing support to the existence of the Berry phase in the vortex action. However, it was not a *derivation* in the rigorous sense of the word. We discuss a new line of thinking that can capture the vortex action, including the Berry phase, within the hydrodynamic framework. To this end, we revisit the GP theory.

The GP Lagrangian in free space ($\hbar \equiv 1$)

$$L = i\psi^*\partial_t\psi - \frac{1}{2m}(\nabla\psi^* \cdot \nabla\psi) - \frac{g}{2}|\psi|^4 \tag{4.29}$$

undergoes a hydrodynamic "surgery" when one writes the complex scalar field $\psi = \sqrt{\rho}e^{i\theta}$. The hydrodynamic fields were deemed smooth, differentiable and so on, so that one could expand the hydrodynamic Lagrangian to second order in the smooth density fluctuation $\delta\rho$ and the phase fluctuation θ. Such an assumption is not likely to hold very well if a vortex is present, since the phase field for the vortex θ_v must have the singular property

$$\nabla_2 \times \nabla_2\theta_v = 2\pi q_v\delta^2(\mathbf{r} - \mathbf{R}_v). \tag{4.30}$$

This is a fact that could (and should) have been accounted for more accurately in passing to the hydrodynamic description. This time we try to do a better job by writing the phase field as the sum of smooth (θ) and singular (θ_v) parts,

$$\psi = \sqrt{\rho}e^{i\theta}e^{i\theta_v}. \tag{4.31}$$

To simplify later notations we will write the vortex phase field by $\chi_v = e^{i\theta_v}$. The singular property (4.30) in terms of χ_v becomes

$$\nabla_2 \times [\chi_v^* \nabla_2 \chi_v] = 2\pi i q_v \delta^2(\mathbf{r} - \mathbf{R}). \tag{4.32}$$

The hydrodynamic Lagrangian derived in (4.3) becomes more complicated now that the vortex phase field has been included,

$$L = -\rho(\dot{\theta} - i\chi_v^* \partial_t \chi_v) - \frac{(\nabla\rho)^2}{8m\rho} - \frac{g}{2}(\rho - \rho_0)^2 - \frac{\rho_0}{2m}(\nabla\theta - i\chi_v^* \nabla\chi_v)^2. \tag{4.33}$$

The density ρ has been replaced by its mean value ρ_0 in the last term.

One invariably notes the asymmetry in the hydrodynamic Lagrangian in the way the temporal part $\chi_v^* \partial_t \chi_v$ appears linearly but the spatial part $\chi_v^* \nabla \chi_v$ appears quadratically. This asymmetry can be remedied by the trick known as the Hubbard–Stratonovich transformation. This is a technique to rewrite

$$-\frac{\rho_0}{2m}(\nabla\theta - i\chi_v^* \nabla\chi_v)^2 \Rightarrow \frac{1}{2} \cdot \frac{2m}{\rho_0}\mathbf{J}^2 - \mathbf{J} \cdot (\nabla\theta - i\chi_v^* \nabla\chi_v), \tag{4.34}$$

so that the spatial gradient $\chi_v^* \nabla \chi_v$ also appears linearly. The original form is recovered by completing the square and integrating out the Hubbard-Stratonovich field $\mathbf{J}$.

The new hydrodynamic action reads

$$\begin{aligned} L = \; & \frac{1}{2} \cdot \frac{2m}{\rho_0}\mathbf{J}^2 - \frac{(\nabla\rho)^2}{8m\rho} - \frac{g}{2}(\rho - \rho_0)^2 \\ & - \rho(\dot{\theta} - i\chi_v^* \partial_t \chi_v) - \mathbf{J} \cdot (\nabla\theta - i\chi_v^* \nabla\chi_v). \end{aligned} \tag{4.35}$$

Both terms in the second line share a similar structure and contain only first derivatives. In fact, one can express them nicely as a single expression by introducing the three-current $J_\alpha = (\rho, \mathbf{J})$:

$$-J_\alpha(\partial_\alpha\theta - i\chi_v^* \partial_\alpha \chi_v). \tag{4.36}$$

The first of these terms, $-J_\alpha \partial_\alpha \theta$, can be integrated by parts to yield a term $\theta \partial_\alpha J_\alpha$ in the action. Since this is the only term in the action that contains θ, i.e., $e^{i\int \theta(\partial_\alpha J_\alpha)}$, integrating out θ in the path integral must give $\delta(\partial_\alpha J_\alpha)$. This is nothing but the constraint on the three-current:

$$\partial_\alpha J_\alpha = 0. \tag{4.37}$$

The left-hand side is like the divergence of a vector field in three dimensions, and from vector analysis we know that this kind of constraint is solved by rewriting the three-current as the curl of another vector field:

$$J_\alpha = \frac{1}{2\pi}\varepsilon_{\alpha\beta\gamma}\partial_\beta a_\gamma, \tag{4.38}$$

where the Greek indices α, β, γ run over the spacetime coordinates t, x, y.

Instead of using J_α, which is a constrained field, we can use the unconstrained field a_α to express the hydrodynamic Lagrangian. Substituting (4.38) back into the action (4.36) gives

$$iJ_\mu[\chi_v^*\partial_\alpha\chi_v] = \frac{i}{2\pi}\left[\varepsilon_{\alpha\beta\gamma}\partial_\beta a_\gamma\right]\left[\chi_v^*\partial_\alpha\chi_v\right] = \frac{i}{2\pi}\varepsilon_{\alpha\beta\gamma}a_\alpha\partial_\beta[\chi_v^*\partial_\gamma\chi_v], \tag{4.39}$$

where the final form comes from integrating the second term by parts.

At first sight, the physical meaning of the new action (4.39) might seem obscure. The best way to unravel it by looking at the expression for the $\alpha = t$ component, where we find

$$\frac{i}{2\pi}a_t\nabla_2\times[\chi_v^*\nabla_2\chi_v] = -q_v a_t\delta^2(\mathbf{r}-\mathbf{R}_v). \tag{4.40}$$

The delta function $q_v\delta^2(\mathbf{r}-\mathbf{R}_v)$ has the obvious interpretation of the vortex charge density. This association can be generalized to include other components in the coupling (4.39). Introducing the vortex three-current

$$J_\alpha^v = \frac{-i}{2\pi}\varepsilon_{\alpha\beta\gamma}\partial_\beta[\chi_v^*\partial_\gamma\chi_v], \tag{4.41}$$

we can use it to rewrite the hydrodynamic Lagrangian (4.35) as

$$L = \frac{1}{2}\cdot\frac{2m}{\rho_0}\mathbf{J}^2 - \frac{(\nabla J_0)^2}{8m\rho_0} - \frac{g}{2}(J_0-\rho_0)^2 - a_\alpha J_\alpha^v. \tag{4.42}$$

The $-a_\alpha J_\alpha^v$ coupling is analogous to the (matter)-(gauge field) coupling in ordinary electrodynamics.

We have obviously come a long way from the pristine GP Lagrangian (4.29) to the form given above. In principle, all the steps of the transformation have been exact. The complex scalar field ψ in the original GP action has been completely superseded by new degrees of freedom: the vortex three-current J_α^v and a second, Hubbard–Stratonovich three-current $J_\alpha = (J_0, \mathbf{J})$. Although the physical significance of $\mathbf{J}$ has never been very clear since it had been introduced purely as a matter of mathematical convenience, fortunately, we can "integrate out" all such Hubbard–Stratonovich fields from the Lagrangian.

First of all, the interaction constraint $g(J_0-\rho_0)^2/2$ forces J_0 to equal the average boson density ρ_0:

$$J_0 = \frac{1}{2\pi}\nabla_2\times\mathbf{a} = \rho_0. \tag{4.43}$$

In other words, the vector field **a** can be solved as if it were a vector potential for the uniform magnetic field $B = 2\pi\rho_0$:

$$\nabla \times \mathbf{a} = 2\pi\rho_0 \;\Rightarrow\; \mathbf{a} = \pi\rho_0(-y, x). \tag{4.44}$$

Next, the vortex three-current of a single vortex localized at $\mathbf{R}(t)$ is given by

$$J^v_\mu = q_v\big(1, -\dot{X}, -\dot{Y}\big)\delta^2(\mathbf{r}-\mathbf{R}), \tag{4.45}$$

and the (matter)-(gauge field) coupling appearing at the end of the hydrodynamic action (4.42) is reduced to (restoring $\hbar$)

$$-\int d^2\mathbf{r}\; a_\alpha J^v_\alpha = -a_t\rho_v + \frac{1}{2}hq_v\rho_0\hat{z}\cdot(\mathbf{R}\times\dot{\mathbf{R}}). \tag{4.46}$$

The second term on the right is precisely the Berry phase action obtained by the Wu-Haldane-Ao-Thouless argument!

The only remaining task is to remove the a_t field from the action by somehow integrating it out. Using our knowledge of (a_x, a_y), we can write

$$\mathbf{J} = \frac{1}{2\pi}(\partial_y a_t - \partial_t a_y, \partial_t a_x - \partial_x a_t) = \frac{1}{2\pi}(\partial_y a_t, -\partial_x a_t), \tag{4.47}$$

and replace $\mathbf{J}^2$ in the Lagrangian by $(\nabla a_t)^2/4\pi^2$. The Lagrangian (4.42) is then reduced to

$$L = \frac{1}{2}a_t\left(-\frac{2m}{\rho_0}\nabla^2\right)a_t - a_t\rho_v + \frac{1}{2}hq_v\rho_0\hat{z}\cdot(\mathbf{R}\times\dot{\mathbf{R}}), \tag{4.48}$$

which is quadratic in a_t. Integrating out the a_t field from the action by performing the Gaussian integration yields an effective Lagrangian expressed entirely in terms of the vortex density ρ_v:

$$L_v = -\frac{1}{2}\frac{\rho_0}{2m}\int d^2\mathbf{r}d^2\mathbf{r}'\;\rho_v(\mathbf{r})\left(\ln|\mathbf{r}-\mathbf{r}'|\right)\rho_v(\mathbf{r}'). \tag{4.49}$$

A collection of point-like vortices is described by the vortex density function $\rho_v = \sum_i q_i\delta^2(\mathbf{r}-\mathbf{R}_i)$. A Berry phase Lagrangian can be simultaneously assigned to each vortex. In this manner, we arrive at the multi-vortex effective Lagrangian

$$L_v = \frac{1}{2}h\rho_0\sum_i q_i\hat{z}\cdot(\mathbf{R}_i\times\dot{\mathbf{R}}_i) - \frac{\rho_0}{2m}\sum_{i<j}q_iq_j\ln|\mathbf{R}_i-\mathbf{R}_j|. \tag{4.50}$$

The inter-vortex interaction turns out as logarithmic.

In hindsight, the GP action contained sufficient ingredient to provide the Berry phase dynamics of the vortex. However, extracting that information required a special trick called the duality transformation, which mapped the whole boson problem to one in terms of vortices. Vortices are considered the fundamental particles of the dual theory.

4.2 Effective Equation for Skyrmion Motion

In this section, we discuss several different ways to deduce the effective dynamics of a skyrmion, taking the view that it can be considered as a point-particle object moving in a ferromagnetic background.

4.2.1 *Spin-Action Approach*

The effective action for the skyrmion in terms of its coordinates $\mathbf{R} = (X, Y)$ can be derived from the microscopic spin-action. In Chap. 1, the single-spin Lagrangian we derived was [cf. (1.19)]

$$L = -\hbar S(1 - \cos\theta)\dot{\phi} - \hbar S\gamma H.$$

In a straightforward generalization of the single-spin action, the continuum action for two-dimensional ferromagnets is

$$\begin{aligned} S &= -\frac{\hbar S}{a^2}\int d^2\mathbf{r}dt(1 - \cos\theta)\partial_t\phi - \int dt E[\mathbf{n}] \\ E[\mathbf{n}] &= \frac{\hbar S\gamma}{a^2}\int d^2\mathbf{r}\; H[\mathbf{n}, \partial_\mu\mathbf{n}, \mathbf{r}]. \end{aligned} \tag{4.51}$$

The transition from the lattice to the continuum action necessitates the appearance in the prefactor of a denominator with dimensions of $(\text{length})^2$, which is provided by the square of the lattice spacing a. Variation of the action (4.51) gives

$$\delta S = \frac{\hbar S}{a^2}\int d^2\mathbf{r}dt\left(\mathbf{n}\times\dot{\mathbf{n}} - \gamma\frac{\delta H}{\delta\mathbf{n}}\right)\cdot\delta\mathbf{n}, \tag{4.52}$$

and the physical spin dynamics are retrieved when the expression inside the parenthesis vanishes:

$$\mathbf{n}\times\dot{\mathbf{n}} - \gamma\frac{\delta H}{\delta\mathbf{n}} = 0. \tag{4.53}$$

This is none other than the continuum Landau–Lifshitz equation. We can introduce the "force" vector as $\mathbf{f} = -\delta H/\delta \mathbf{n}$ to rewrite the LL equation in the familiar form

$$\dot{\mathbf{n}} = \gamma \mathbf{n} \times \mathbf{f}. \tag{4.54}$$

Specializing to the situation in which the only source of motion is that of the skyrmion coordinates $\mathbf{R}$ allows us to give the following prescription for the spin configuration

$$\mathbf{n}(\mathbf{r}, t) = \mathbf{n}(\mathbf{r} - \mathbf{R}(t)). \tag{4.55}$$

Under this assumption, variations in the spin configuration $\delta\mathbf{n}$ can only be a consequence of the displacement of the skyrmion position $\mathbf{R}(t) \to \mathbf{R}(t) + \delta\mathbf{R}(t)$. As a result, we have

$$\delta\mathbf{n} = -(\delta X \partial_x \mathbf{n} + \delta Y \partial_y \mathbf{n}), \quad \dot{\mathbf{n}} = -(\dot{X}\partial_x \mathbf{n} + \dot{Y}\partial_y \mathbf{n}). \tag{4.56}$$

Inserting these terms back into the geometric part of the variation (4.52) yields

$$\begin{aligned} \delta S_{\mathrm{B}} &= \frac{\hbar S}{a^2}\left(\int d^2\mathbf{r}\, \mathbf{n}\cdot(\partial_x \mathbf{n} \times \partial_y \mathbf{n})\right)\int dt(\dot{X}\delta Y - \dot{Y}\delta X) \\ &= \frac{2hSQ_s}{a^2}\int dt(\dot{X}\delta Y - \dot{Y}\delta X). \end{aligned} \tag{4.57}$$

The skyrmion charge Q_s appears as a prefactor in this skyrmion action, as the vortex charge q_v appeared as a prefactor in its action (4.22). We see that δS_{B} has been factorized as the integral over the space $\mathbf{r}$ yielding the skyrmion charge, and an integral over time, which depends only on the skyrmion coordinates $\mathbf{R}(t)$ and its displacement $\delta\mathbf{R}(t)$. As in the derivation of the vortex action, we ask what kind of point-particle action yields the same variation as (4.57). As in the vortex case before, it is easy to guess the answer,

$$S_s = -\frac{hSQ_s}{a^2}\int dt(X\dot{Y} - Y\dot{X}), \quad S_v = \frac{hq_v}{2a^2}\int dt(X\dot{Y} - Y\dot{X}). \tag{4.58}$$

For comparison, the corresponding vortex action is also displayed after rewriting the superfluid density ρ_0 as $1/a^2$. The commutator for the guiding center coordinates of the vortex derived in (4.28) finds a skyrmion analogue

$$[X_g, Y_g] = \frac{ia^2}{4\pi SQ_s}. \tag{4.59}$$

In Chap. 1, we argued that the squared time derivative term $(\dot{\mathbf{n}})^2$ had to be absent from the spin action in order to preserve the norm. This restriction implies that a "bare mass" term such as $M\dot{\mathbf{R}}^2$ is also forbidden in the action for a skyrmion. Recall that the bare hydrodynamic action for the superfluid (4.3) was linear in the time derivative $\dot{\theta}$, and to obtain the vortex mass, a squared term had to be generated through the coupling of $\dot{\theta}$ to the density fluctuation $\delta\rho$. In fact, a similar mechanism can be invoked to generate the mass of a skyrmion, but we will defer discussion of the skyrmion mass to Sect. 6.4, where it can be shown that integrating out some deformation modes of the skyrmion generates the desired $\dot{\mathbf{R}}^2$ term. For now, the skyrmion mass M_s will be treated as a phenomenological constant.

Applying the skyrmion ansatz $\mathbf{n}(\mathbf{r}, t) = \mathbf{n}(\mathbf{r} - \mathbf{R}(t))$ in the energy part of the action (4.51) gives an effective potential energy

$$\begin{aligned} V(\mathbf{R}) &= \frac{\hbar S\gamma}{a^2} \int d^2\mathbf{r}\; H[\mathbf{n}(\mathbf{r}-\mathbf{R}), \partial_\mu \mathbf{n}(\mathbf{r}-\mathbf{R}), \mathbf{r}] \\ &= \frac{\hbar S\gamma}{a^2} \int d^2\mathbf{r}\; H[\mathbf{n}(\mathbf{r}), \partial_\mu \mathbf{n}(\mathbf{r}), \mathbf{r}+\mathbf{R}]. \end{aligned} \tag{4.60}$$

The first line of the energy functional is written with the assumption that H depends on $\mathbf{r}$ explicitly. Without such dependence, the integral on the right should become independent of $\mathbf{R}$, and incapable of serving as a potential energy function for the skyrmion. This is not surprising since, in a completely homogeneous environment, one must find the same energy no matter where the skyrmion is placed. However, if the energy functional does break translational symmetry by virtue of some impurity potential or inhomogeneous Zeeman field, the resulting $\mathbf{R}$-dependent function $V(\mathbf{R})$ can be viewed as the potential energy of the skyrmion. Collecting all the terms derived so far, one arrives at the effective skyrmion Lagrangian

$$L_s = \frac{1}{2} M_s \dot{\mathbf{R}}^2 - \frac{1}{2} G\hat{z} \cdot (\mathbf{R} \times \dot{\mathbf{R}}) - V(\mathbf{R}). \tag{4.61}$$

The constant

$$G = 2hSQ_s/a^2 \tag{4.62}$$

is known as the gyrotropic constant and appears often in the discussion of skyrmion dynamics. The skyrmion equation of motion following from the action is

$$M_s \ddot{\mathbf{R}} = -\frac{\partial V}{\partial \mathbf{R}} + G\hat{z} \times \dot{\mathbf{R}}. \tag{4.63}$$

This is none other than Lorentz's equation, with the first and second terms on the right-hand side acting as the electric and magnetic forces, respectively. For vanishing skyrmion mass the two forces must balance out, producing the drift motion

$$G\dot{\mathbf{R}} = \hat{z} \times \left(-\frac{\partial V}{\partial \mathbf{R}} \right). \tag{4.64}$$

The rigid-skyrmion approximation $\mathbf{n}(\mathbf{r}, t) = \mathbf{r}(\mathbf{r} - \mathbf{R}(t))$ that we used to derive the Berry phase action (4.57) for a skyrmion in a ferromagnet was adopted by Stone [9] early on to address the skyrmion dynamics in quantum Hall ferromagnets. Later, the dynamics of skyrmions in a chiral ferromagnet was developed along similar lines of reasoning in Ref. [10]. Much earlier than both, Thiele had adopted the rigid-body ansatz to derive the dynamics of a magnetic vortex (which is really a skyrmion in a ferromagnet) from the Landau–Lifshitz equation [11]. Thiele's derivation, which was restricted to the steady-state motion of the skyrmion, was generalized to treat arbitrary motion in Refs. [12, 13]. The derivation of skyrmion dynamics from a generalized version of Thiele's approach is the next topic of discussion.

4.2.2 Landau–Lifshitz–Gilbert Equation Approach

The action-based derivation of the skyrmion dynamics fails to capture the effect of dissipation on the skyrmion motion. On the other hand, the Landau–Lifshitz equation is easily supplemented with the damping term, leading to the Landau–Lifshitz–Gilbert (LLG) equation

$$\dot{\mathbf{n}} = \gamma \mathbf{n} \times \mathbf{f} - \alpha \mathbf{n} \times \dot{\mathbf{n}}. \tag{4.65}$$

For practical magnetic materials, the magnitude of the damping parameter α (Gilbert constant) is typically less than 0.1. In fact, the LLG equation can be inverted exactly to give

$$\dot{\mathbf{n}} = \frac{\gamma}{1+\alpha^2}[\mathbf{n} \times \mathbf{f} - \alpha \mathbf{n} \times (\mathbf{n} \times \mathbf{f})], \tag{4.66}$$

with no time derivative terms appearing on the right-hand side. The LLG equation in this latter form is widely used in simulating the spin dynamics of a magnet. In most cases found in the literature, the LLG equation is written in terms of the magnetization vector rather than the spin vector as we do here. The magnetic form can be recovered by taking $\mathbf{n} \to -\mathbf{n}$ everywhere in (4.65) and (4.66).

We are familiar with the damping term in Newtonian dynamics. It is usually linear in the particle's velocity, and the total energy tends to decrease in proportion to the square of the velocity. The Gilbert term in the spin dynamics, on the other hand, does not immediately come across as a dissipative term. One can, in fact, prove that the Gilbert term $\alpha \mathbf{n} \times \dot{\mathbf{n}}$ in (4.65) is responsible for damping. The total energy in arbitrary dimension d is

$$E[\mathbf{n}] = \frac{\hbar S \gamma}{a^d} \int d^d\mathbf{r} \; H[\mathbf{n}, \partial_\mu \mathbf{n}, \mathbf{r}], \tag{4.67}$$

and working out its rate of change gives

$$\dot{E} = \frac{\hbar S\gamma}{a^d} \int d^d\mathbf{r}\, \frac{\delta H}{\delta \mathbf{n}} \cdot \dot{\mathbf{n}} = -\frac{\hbar S\gamma}{a^d} \int d^d\mathbf{r}\, \mathbf{f} \cdot \dot{\mathbf{n}}. \tag{4.68}$$

In place of $\dot{\mathbf{n}}$, one can insert the right-hand side of the LLG equation (4.65) to get

$$\dot{E} = -\alpha \frac{\hbar S\gamma}{a^d} \int d^d\mathbf{r}\, (\mathbf{f} \times \mathbf{n}) \cdot \dot{\mathbf{n}}. \tag{4.69}$$

Invoking the LLG equation one more time to replace $\mathbf{f} \times \mathbf{n}$ by $(\dot{\mathbf{n}} - \alpha\mathbf{n} \times \dot{\mathbf{n}})/\gamma$, we finally obtain

$$\dot{E} = -\frac{\alpha\hbar S}{a^d} \int d^d\mathbf{r}\, \dot{\mathbf{n}}^2 < 0. \tag{4.70}$$

The energy is a strictly decreasing function of time indeed under the LLG dynamics.

As in the action-based derivation of skyrmion dynamics, we adopt a rigid skyrmion ansatz and write $\mathbf{n}(\mathbf{r}, t) = \mathbf{n}(\mathbf{r} - \mathbf{R}(t))$. Feeding $\dot{\mathbf{n}} = -\dot{X}\partial_x\mathbf{n} - \dot{Y}\partial_y\mathbf{n}$ on either side of the LLG equation (4.65) and taking the cross product with $\mathbf{n}$ gives

$$\dot{X}(\mathbf{n} \times \partial_x\mathbf{n}) + \dot{Y}(\mathbf{n} \times \partial_y\mathbf{n}) = \gamma\mathbf{f} + \alpha\dot{X}\partial_x\mathbf{n} + \alpha\dot{Y}\partial_y\mathbf{n}. \tag{4.71}$$

In arriving at this expression, we made the assumption that $\mathbf{f}$ has components in the transverse direction only, i.e., $\mathbf{n} \cdot \mathbf{f} = 0$. This assumption is justified since the force term $\mathbf{f}$ enters in the LLG equation as a cross product $\mathbf{n} \times \mathbf{f}$, and any longitudinal portion in $\mathbf{f}$ would automatically drop out. As a result, one might as well regard $\mathbf{f}$ as entirely transverse in the first place.

Taking the inner product of both sides with $\partial_x\mathbf{n}$ and $\partial_y\mathbf{n}$, respectively, and integrating over the whole two-dimensional space, one arrives at a pair of equations

$$\begin{aligned} 4\pi Q_s\dot{X} &= \gamma \int d^2\mathbf{r}\, (\partial_y\mathbf{n} \cdot \mathbf{f}) + 4\pi\alpha\eta\dot{Y}, \\ 4\pi Q_s\dot{Y} &= -\gamma \int d^2\mathbf{r}\, (\partial_x\mathbf{n} \cdot \mathbf{f}) - 4\pi\alpha\eta\dot{X}. \end{aligned} \tag{4.72}$$

In the above, we have used the relation $\int d^2\mathbf{r}(\partial_x\mathbf{n}) \cdot (\partial_y\mathbf{n}) = 0$, assuming a circularly symmetric skyrmion profile, and introduced the shape factor

$$\eta = \frac{1}{4\pi} \int d^2\mathbf{r}\, (\partial_x\mathbf{n})^2 = \frac{1}{4\pi} \int d^2\mathbf{r}\, (\partial_y\mathbf{n})^2. \tag{4.73}$$

Further manipulation of (4.72) is possible with the following substitution:

$$\begin{aligned} \int d^2\mathbf{r}\, (\partial_x\mathbf{n} \cdot \mathbf{f}) &= -\int d^2\mathbf{r}\, \partial_x\mathbf{n} \cdot \frac{\delta H}{\delta \mathbf{n}} \\ &= -\int d^2\mathbf{r}\, \partial_x H + \int d^2\mathbf{r}\, \text{“}\partial_x\text{”} H. \end{aligned} \tag{4.74}$$

Here we must carefully go over the meaning of the two partial derivatives. The first one in the second line acts both on the indirect x-dependence through $\mathbf{n}$ and $\partial_\mu \mathbf{n}$, and a direct x-dependence. The partial derivatives appearing in the first line, on the other hand, are only considering the indirect dependences on x, so that the direct partial derivative "∂_x" had to be subtracted out, as in the last term of the second line. As a total derivative, the first integral in the second line vanishes, and the second integral, after some careful reshuffling of the variables, becomes

$$\begin{aligned}
&\int d^2\mathbf{r}\ \text{``}\partial_x\text{''} H[\mathbf{n}(\mathbf{r}-\mathbf{R}), \partial_\mu \mathbf{n}(\mathbf{r}-\mathbf{R}), \mathbf{r}] \\
&= \int d^2\mathbf{r}\ \text{``}\partial_x\text{''} H[\mathbf{n}(\mathbf{r}), \partial_\mu \mathbf{n}(\mathbf{r}), \mathbf{r}+\mathbf{R}] \\
&= \int d^2\mathbf{r}\ \partial_X H[\mathbf{n}(\mathbf{r}), \partial_\mu \mathbf{n}(\mathbf{r}), \mathbf{r}+\mathbf{R}] \\
&= \frac{a^2}{\hbar S \gamma} \partial_X V(\mathbf{R}). \qquad (4.75)
\end{aligned}$$

The potential energy $V(\mathbf{R})$ of the skyrmion in the last line comes from the earlier definition (5.11). Now, (4.72) can be summed up as a single vectorial equation,

$$G\dot{\mathbf{R}} = \left(\frac{\partial V}{\partial \mathbf{R}} + G\frac{\alpha\eta}{Q_s}\dot{\mathbf{R}} \right) \times \hat{z}. \qquad (4.76)$$

Including the phenomenological mass term yields the Newtonian equation of motion,

$$M_s\ddot{\mathbf{R}} = -\frac{\partial V}{\partial \mathbf{R}} + G\left(\hat{z} \times \dot{\mathbf{R}} - \frac{\alpha\eta}{Q_s}\dot{\mathbf{R}} \right), \qquad (4.77)$$

which differs from (4.63) by the inclusion of the additional damping term. We can define a skyrmion Hall angle $\tan\theta = \alpha\eta/Q_s$ from this equation. As in the electronic Hall effect, the sign of the Hall angle depends on the charge, which in this case is the topological charge Q_s.

4.2.3 Alternative Approach

One final derivation of the skyrmion dynamics will be given. The discussion in this section follows the field-theoretical work of Papanicolaou and Tomaras [14], which was developed well before the experimental discovery of chiral skyrmions. In this approach, one works in the context of the temporal dynamics of the emergent magnetic field

$$b_\mu = \frac{1}{4}\varepsilon_{\mu\nu\rho}\mathbf{n} \cdot (\partial_\nu \mathbf{n} \times \partial_\rho \mathbf{n}) = \varepsilon_{\mu\nu\rho}\partial_\nu a_\rho. \qquad (4.78)$$

The Greek letters μ, ν, ρ refer to the three components of the space index x, y, z. The obvious strategy here is that if we know how b_μ evolves over time, we will also be able to learn the dynamics of the integrated object that we are calling the skyrmion.

Invoking the Landau–Lifshitz equation $\dot{\mathbf{n}} = \gamma \mathbf{n} \times \mathbf{f}$ (no damping term), one can show after some algebra that the temporal dynamics of b_μ are governed by the equation

$$\begin{aligned} \partial_t b_\mu &= -\frac{\gamma}{2}\varepsilon_{\mu\nu\rho}\partial_\rho \mathbf{n} \cdot \partial_\nu \mathbf{f} = -\frac{\gamma}{2}\varepsilon_{\mu\nu\rho}\partial_\nu(\partial_\rho \mathbf{n} \cdot \mathbf{f}) \\ &= -\partial_\nu J_{\mu\nu}. \end{aligned} \tag{4.79}$$

This one has the appearance of a continuity equation of sort, reflecting the conservation of the flux $\int d^3\mathbf{r}\; b_\mu$ regardless of the details of the spin Hamiltonian. The ultimate purpose of deriving the topological continuity equation is to see whether the skyrmion equation of motion can be deduced from it.

In two dimensions, a reasonable definition for the skyrmion center coordinates is

$$\mathbf{R} = (X, Y) \equiv \frac{\int d^2\mathbf{r}\; \mathbf{r} b(\mathbf{r})}{\int d^2\mathbf{r}\; b(\mathbf{r})}, \tag{4.80}$$

where we dropped the third index of b_μ to limit our focus to the two-dimensional case. As a topological number, the denominator is independent of time and equals $2\pi Q_s$. The continuity equation derived in (4.79) simplifies to a single equation

$$\partial_t b + \nabla \cdot \mathbf{J} = 0, \tag{4.81}$$

where the two-dimensional topological current $\mathbf{J}$ is defined as $\mathbf{J} = (J_{31}, J_{32})$. Taking the time derivative of $\mathbf{R}$ yields

$$2\pi Q_s \dot{\mathbf{R}} = \int d^2\mathbf{r}\; \mathbf{r}\dot{b} = -\int d^2\mathbf{r}\; \mathbf{r}(\nabla \cdot \mathbf{J}) = \int d^2\mathbf{r}\; \mathbf{J}. \tag{4.82}$$

It turns out, according to this equation, that the skyrmion dynamics depends on the integral of the topological current $\int d^2\mathbf{r}\; \mathbf{J}$.

In (4.79) the topological current $J_{\mu\nu}$ was defined as

$$J_{\mu\nu} = \frac{\gamma}{2}\varepsilon_{\mu\nu\rho}\partial_\rho \mathbf{n} \cdot \mathbf{f}, \tag{4.83}$$

where the vector $\mathbf{f}$ was defined as the variational derivative $\mathbf{f} = -\delta H/\delta \mathbf{n}$. When the energy density functional H does not have any explicit coordinate dependence, $H = H[\mathbf{n}, \partial_\mu \mathbf{n}]$, one can also show

$$\mathbf{f} = -\frac{\delta H}{\delta \mathbf{n}} = \partial_\mu \left[\frac{\partial H}{\partial(\partial_\mu \mathbf{n})}\right] - \frac{\partial H}{\partial \mathbf{n}}. \tag{4.84}$$

Direct substitution of **f** and a few steps of partial derivative algebra gives the expression

$$J_{\mu\nu} = \frac{\gamma}{2}\varepsilon_{\mu\nu\rho}\partial_\rho \mathbf{n}\cdot\mathbf{f} = \frac{\gamma}{2}\varepsilon_{\mu\nu\rho}\partial_\lambda\left[\frac{\partial H}{\partial(\partial_\lambda \mathbf{n})}\cdot\partial_\rho\mathbf{n} - \delta_{\rho\lambda}H\right]. \tag{4.85}$$

We see that, apparently, $J_{\mu\nu}$ is the total derivative of some other quantity

$$T_{\lambda\rho} = \frac{\partial H}{\partial(\partial_\lambda \mathbf{n})}\cdot\partial_\rho\mathbf{n} - \delta_{\lambda\rho}H, \tag{4.86}$$

which is known as the energy-momentum tensor in field theory. Since $J_{\mu\nu} = (\gamma/2)\varepsilon_{\mu\nu\rho}\partial_\lambda T_{\lambda\rho}$, the space integral $\int d^3\mathbf{r}\ J_{\mu\nu}$ vanishes when appropriate boundary conditions on the tensor $T_{\lambda\rho}$ are imposed at spatial infinity. As a result, the skyrmion center coordinate **R** remains stationary. This is easy to understand once we draw on the analogy to the guiding center dynamics of electrons in a quantizing magnetic field. In that context, the electron executes cyclotron motion about an immobile guiding center, and the guiding center becomes mobile only when an electric field is applied, executing a drift motion governed by $\dot{\mathbf{R}} = \mathbf{E}\times\mathbf{B}/\mathbf{B}\cdot\mathbf{B}$. So far in our discussion, there is nothing analogous to the electric field in our formulation of skyrmion dynamics.

Such an analogue to the electric field arises from including an explicit **r** dependence in the Hamiltonian

$$H[\mathbf{n}, \partial_\mu\mathbf{n}] \rightarrow H[\mathbf{n}, \partial_\mu\mathbf{n}, \mathbf{r}]. \tag{4.87}$$

Let us consider the changes that might be brought about by this inhomogeneity. After some thought, one realizes that rewriting $J_{\mu\nu}$ purely in terms of the field tensor, as in (4.85), is only possible for the homogeneous case. Instead, allowing for the explicit coordinate dependence of H and using (4.83) and (4.84), one finds

$$J_{\mu\nu} = \frac{\gamma}{2}\varepsilon_{\mu\nu\rho}\partial_\lambda T_{\lambda\rho} + \frac{\gamma}{2}\varepsilon_{\mu\nu\rho}\text{``}\partial_\rho\text{''}H. \tag{4.88}$$

As was mentioned earlier, the partial derivative appearing in quotation marks above only acts on the explicit coordinate dependence of $H[\mathbf{n}, \partial_\mu\mathbf{n}, \mathbf{r}]$, and not through the implicit coordinate dependence of $\mathbf{n}(\mathbf{r})$ or $\partial_\mu\mathbf{n}(\mathbf{r})$.

Since we already know that the total derivative part of the current tensor $J_{\mu\nu}$ fails to produce the required dynamics for **R**, we focus on the extra inhomogeneous term in (4.88). In particular, the two-dimensional topological current (J_{31}, J_{32}) contains an inhomogeneous contribution

$$\mathbf{J}_{\text{inh.}} = \frac{\gamma}{2}(``\partial_y" H, -``\partial_x" H) = -\frac{\gamma}{2}\hat{z} \times ``\nabla_{\mathbf{r}}" H. \quad (4.89)$$

The skyrmion dynamics resulting from (4.82) are then governed by

$$\begin{aligned} 4\pi Q_s \dot{\mathbf{R}} &= -\gamma \hat{z} \times \int d^2\mathbf{r}\, ``\nabla_{\mathbf{r}}" H \\ &= -\gamma \hat{z} \times \frac{\partial}{\partial \mathbf{R}} \int d^2\mathbf{r}\, H[\mathbf{r}, \partial_\mu \mathbf{n}, \mathbf{r}+\mathbf{R}] \\ &= -\frac{a^2}{\hbar S} \dot{\hat{z}} \times \frac{\partial V}{\partial \mathbf{R}}, \end{aligned} \quad (4.90)$$

where the final line comes from inserting (4.75). We see that the massless limit of the skyrmion equation of motion has been recovered.

Other than providing us with a seemingly redundant derivation of the skyrmion dynamics, the topological continuity equation (4.79) bears an interesting interpretation. Writing the Landau-Lifshitz equation as $\dot{\mathbf{n}} \times \mathbf{n} = \gamma \mathbf{f}$, the topological current

$$\begin{aligned} J_{\mu\nu} &= \frac{\gamma}{2} \varepsilon_{\mu\nu\rho} \partial_\rho \mathbf{n} \cdot \mathbf{f} \\ &= \frac{1}{2} \varepsilon_{\mu\nu\rho} \mathbf{n} \cdot (\partial_\rho \mathbf{n} \times \partial_t \mathbf{n}) = -\varepsilon_{\mu\nu\rho} e_\rho \end{aligned} \quad (4.91)$$

can be cast as the emergent electric field $e_\rho = \frac{1}{2}\mathbf{n} \cdot (\partial_t \mathbf{n} \times \partial_\rho \mathbf{n}) = \partial_t a_\rho - \partial_\rho a_t$. The topological continuity equation (4.79) then becomes

$$\partial_t \mathbf{b} = \nabla \times \mathbf{e}, \quad (4.92)$$

which is the familiar Faraday's law for emergent electromagnetic fields $(\mathbf{b}, \mathbf{e})$. We conclude the following equivalence

$$\begin{aligned} &(\text{Topological continuity equation}) \\ &= (\text{Faraday's law of emergent electrodynamics}). \end{aligned} \quad (4.93)$$

Earlier, a restricted version of the emergent Faraday's law for two-dimensional space was expressed in (2.74) when we discussed the CP^1 mapping of the O(3) vector field.

When the Gilbert damping is included $\alpha \neq 0$, the topological conservation law (4.79) still holds but with $\gamma \mathbf{f}$ replaced by $\gamma \mathbf{f} - \alpha \mathbf{n} \times \mathbf{f}$. As a consequence, the skyrmion number $Q_\mu = \int q_\mu$ remains conserved even in the presence of damping. On the other hand, the relation expressing the topological current $J_{\mu\nu}$ as the total derivative of the energy-momentum tensor, (4.85), is only valid for $\alpha = 0$.

4.3 Applications of the Skyrmion Equation of Motion

4.3.1 Skyrmions in Nanodisks

As a simple application of the skyrmion equation of motion in (4.77), let us consider a single skyrmion confined to a nanodisk, as shown in Fig. 4.2. The dynamics of the skyrmion are governed by

$$M_s\ddot{\mathbf{R}} = -K\mathbf{R} + G\left(\hat{z}\times\dot{\mathbf{R}} - \frac{\alpha\eta}{Q_s}\dot{\mathbf{R}}\right), \tag{4.94}$$

where the confining potential provides the skyrmion with a harmonic restoring force $-K\mathbf{R}$. This problem can be solved exactly in terms of the complex coordinates $Z = X + iY$, for which it reads

$$M_s\ddot{Z} = -KZ + G\left(1 + i\frac{\alpha\eta}{Q_s}\right)\dot{Z}. \tag{4.95}$$

Two characteristic frequencies follow from the solution, $Z(t) = Z_0e^{i\omega t}$:

$$\omega_{\pm} \approx \pm\sqrt{\frac{K}{M_s} + \left(\frac{G}{2M_s}\right)^2} + \frac{G}{2M_s} + i\alpha\eta\frac{G}{Q_s}, \tag{4.96}$$

where a small value of $\alpha\eta$ was assumed in presenting the final expression. The solutions correspond to two gyration modes; one that rotates in a clockwise fashion, while the second rotates counterclockwise. Both modes share the same damping constant $\alpha\eta G/Q_s = \alpha\eta(2hS/a^2)$, which is independent of the skyrmion charge. Further details regarding the derivation of the resonance modes (4.96) is provided in Ref. [13].

From (4.96), it is clear that the counterclockwise mode frequency differs from that of the clockwise mode by G/M_s. In fact, one of the modes disappears in the

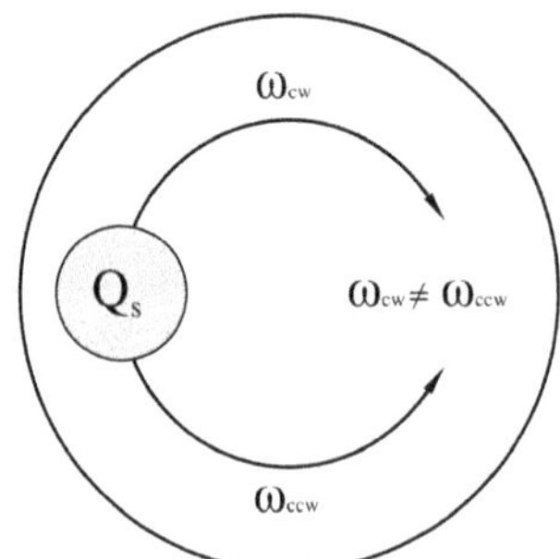

Fig. 4.2 Two gyration modes of a skyrmion ($Q_s \neq 0$) confined in a nanodisk. The frequency difference between the clockwise and counterclockwise modes is due to the finite topological charge

massless limit. As the frequency difference is proportional to G, it is also proportional to the topological charge Q_s. As a useful point of comparison, the first experimental detection of the circulation of a quantized vortex was achieved by Vinen [15] using a thin wire immersed in superfluid ^{4}He having a single quantized vortex line dressing it. The difference of the two vibration modes could be attributed to a nonzero value of G/M_v, where $G \neq 0$ would be due to the vortex charge q_v. Similarly, the first experimental estimation of the skyrmion mass (Ref. [16]) was done by fitting the observed real-time trajectory of a skyrmion to a linear superposition of the two damped modes found in (4.96). The real parts of the frequencies were estimated at 1.00(13) and 1.35(16) GHz, respectively, yielding a skyrmion weight upwards of 8×10^{-22} kg.

4.3.2 Phonon Mode in the Skyrmion Lattice

The point-particle theory of skyrmion dynamics developed thus far can be applied to the dynamics of an array of skyrmions. As discussed at length in Chap. 3, skyrmions form a triangular lattice in thin-film chiral magnets. Viewing the skyrmion lattice as a collection of individually well-defined skyrmions, a multi-skyrmion Lagrangian can be written as

$$L[\{\mathbf{R}_i\}] = \frac{M_s}{2}\sum_i \dot{\mathbf{R}}_i^2 - \frac{G}{2}\sum_i \hat{z}\cdot(\mathbf{R}_i \times \dot{\mathbf{R}}_i) - \frac{K}{2}\sum_i \mathbf{R}_i^2 - \sum_{i<j} V(\mathbf{R}_i - \mathbf{R}_j). \quad (4.97)$$

An inter-skyrmion interaction energy $V(\mathbf{R}_i - \mathbf{R}_j)$ has been introduced. For completeness, we include the same confining potential $K\mathbf{R}_i^2/2$ used in the nanodisk geometry calculation.

The inter-vortex interaction in superfluid is long-ranged, falling off only logarithmically with distance. The vortex-vortex interaction in a superconductor is limited by the screening length, which is a distinct quantity from the inter-vortex distance that is set by the external magnetic field and the magnetic length. In the skyrmion lattice, both the cutoff length of the pairwise skyrmion interaction and the inter-skyrmion distance are set by the same length scale, i.e., the spiral wavelength λ. In a harmonic approximation one writes the pairwise potential as

$$V(\mathbf{R}_i - \mathbf{R}_j) = \frac{V_0}{2R_0^2}(\mathbf{R}_i - \mathbf{R}_j)^2, \quad (4.98)$$

where $\mathbf{R}_i$ stands for the displacement of the skyrmion from its equilibrium position $\mathbf{R}_i^0$, and R_0 is the inter-skyrmion distance in the equilibrium state of the skyrmion lattice. The equation of motion for the collection of skyrmions becomes

$$M_s\ddot{X}_i + KX_i - G\dot{Y}_i + \frac{V_0}{R_0^2}\sum_{j\neq i}(X_i - X_j) = 0$$
$$M_s\ddot{Y}_i + KY_i + G\dot{X}_i + \frac{V_0}{R_0^2}\sum_{j\neq i}(Y_i - Y_j) = 0. \tag{4.99}$$

Introducing a harmonic displacement field $Z_i = X_i + iY_i = Z_0 e^{i\mathbf{k}\cdot\mathbf{R}_i^0 - i\omega_{\mathbf{k}}t}$ yields the dispersion

$$-M_s\omega_{\mathbf{k}}^2 + K + G\omega_{\mathbf{k}} + \frac{V_0}{R_0^2}\sum_{j\neq i}\left(1 - e^{i\mathbf{k}\cdot(\mathbf{R}_j^0 - \mathbf{R}_i^0)}\right) = 0. \tag{4.100}$$

The structure factor appearing on the right-side of this equation in the case of a triangular skyrmion lattice is

$$\frac{2V_0}{R_0^2}\left(3 - \cos(k_x R_0) - \cos\left[\left(\frac{k_x + \sqrt{3}k_y}{2}\right)R_0\right] - \cos\left[\left(\frac{k_x - \sqrt{3}k_y}{2}\right)R_0\right]\right)$$
$$\approx \frac{3}{2}V_0\mathbf{k}^2. \tag{4.101}$$

Solving (4.100) yields two eigenmodes, with frequencies

$$\omega_\pm(\mathbf{k}) = \frac{G}{2M_s} \pm \sqrt{\left(\frac{G}{2M_s}\right)^2 + \frac{K}{M_s} + \frac{3V_0}{2M_s}\mathbf{k}^2}. \tag{4.102}$$

Without the confining potential ($K = 0$) the Lagrangian is invariant under the continuous translation of all the skyrmion coordinates. The Goldstone mode arising from the broken translational symmetry of the skyrmion lattice has the low-frequency dispersion

$$\omega_+(\mathbf{k}) \sim (3V_0/2|G|)\mathbf{k}^2, \tag{4.103}$$

which varies quadratically with the wavevector. Here we are assuming $G < 0$, so that $|\omega_+| < |\omega_-|$. The "optical phonon" mode $\omega_-(\mathbf{k})$ has a gap of size $|G|/M_s$ at $\mathbf{k} = 0$. For general $K \neq 0$, the two gapped $\mathbf{k} = 0$ modes represent the uniform clockwise and counterclockwise rotation of all the skyrmions in the lattice. One of those cyclotron-like motions turns into the Goldstone mode as $K \to 0$.

As an acoustic phonon mode, the quadratic dependence $\omega_-(\mathbf{k}) \sim \mathbf{k}^2$ seems peculiar since ordinary phonons obey a linear dispersion $\omega_{\mathbf{k}} = c_s|\mathbf{k}|$, with the speed of sound c_s. Additionally, two-dimensional crystals typically support two kinds of phonons because there are two possible displacement directions, one longitudinal and one transverse. In the skyrmion lattice, however, those two modes get mixed, giving rise to only one acoustic mode while the other mode becomes gapped. In both its manner of derivation and in its consequence, a close parallel exists between the acoustic phonon mode of the skyrmion lattice and the spin-wave mode of a ferromagnet. In both, the dispersion scales quadratically with the wavevector, and the two fluctuation directions (X_i and Y_i for skyrmions, S_i^x and S_i^y for ferromagnets aligned along the z-direction) are coupled. The similarity is not incidental, but arises from the analogous commutation algebra which, for ferromagnetic spins, is

$$[S_i^x, S_i^y] = i\hbar S_i^z \approx i\hbar \langle S_i^z \rangle. \tag{4.104}$$

The two variables S_i^x, S_i^y act as canonical conjugates. In the skyrmion case, the two guiding center coordinates X and Y form the canonical pair, obeying a similar commutation algebra [cf. (4.59)]

$$[X, Y] = i\frac{a^2}{4\pi S Q_s}.$$

4.3.3 Numerical and Experimental Tests

An extensive numerical simulation study of the low-energy excited modes in the skyrmion crystal phase was first carried out by Mochizuki [17]. In his analysis, the LLG equation was integrated numerically assuming a time-dependent Zeeman field of the form

$$H_Z(t) = -(\mathbf{B} + \mathbf{B}_p(t)) \cdot \mathbf{n}. \tag{4.105}$$

At first $\mathbf{B}_p(t)$ was applied in the form of a pulse, and the subsequent time evolution of the spin vectors was analyzed in the frequency domain in order to obtain the resonance frequencies. Subsequently, a harmonic oscillating field $\mathbf{B}_p(t)$, fixed at one of the resonance frequencies, was applied to excite the individual resonance mode.

Two resonance peaks were predicted for in-plane oscillating magnetic field, corresponding to the clockwise and counterclockwise rotation of the skyrmion center. Only the $\mathbf{k} = 0$ collective modes were discernible in the simulation due to the uniform nature of the applied field. For realistic values of J about 100 GHz, two resonance frequencies were found around 1 GHz, i.e., in the microwave range. These two modes are reminiscent of the modes we found in the nanodisk calculation of Sect. 4.3.1. When the oscillating field was applied along the external field direction, perpendicular to the plane, the simulation results found a single resonance mode identified as the breathing mode of a skyrmion. The breathing mode requires a description beyond the point-particle dynamics of the present chapter, and will be discussed in Sect. 6.4.

On the experimental side, microwave studies on the insulating skyrmion crystal Cu_2OSeO_3 found two resonance modes in the GHz range [18], one for an in-plane oscillating field, and one for an out-of-plane oscillating field. These modes were interpreted as the counterclockwise gyration mode and the breathing mode, respectively. Similar microwave experiments were performed in the metallic compound MnSi and $Fe_{1-x}Co_xSi$ [19]. Both modes of gyration were picked up for MnSi, but only the counterclockwise mode was observed for $Fe_{1-x}Co_xSi$. The weaker intensity of the clockwise mode is responsible for the difficulties of its detection.

As we saw, the phonon calculation for the skyrmion lattice in Sect. 4.3.2 predicted a vanishing gap for one of the $\mathbf{k} = 0$ modes, in apparent disagreement with experiments and numerical calculations. The resolution of this problem comes from the diagonalization of the lattice magnon Hamiltonian, which identified the clockwise, counterclockwise, and breathing modes as the $\mathbf{k} = 0$ excitations in some of the low-lying, but (quite significantly!) not the lowest lying magnon bands [19]. According to the latest analysis [20], they are the $\mathbf{k} = 0$ excitations of the 3rd (counterclockwise), 5th (breathing), and 6th (clockwise) magnon bands.

References

1. Kelvin, L. (Sir Thomson, W.): On vortex atoms. In: Proceedings of the Royal Society of Edinburgh, vol. VI, 1867, pp. 94–105. Reprinted in Phil. Mag., vol. XXXIV, 1867, pp. 15–24
2. Onsager, L.: Suppl. Nuovo Cim. **6**, 249 (1949)
3. Thouless, D.J.: Topological Quantum Numbers in Nonrelativistic Physics. World Scientific, Singapore (1998)
4. Lamb, H.: Hydrodynamics. Cambridge University Press, New York (1975)
5. Feynman, R.P.: Application of quantum mechanics to liquid helium. In: Gorter, C.J. (ed.) Progress in Low-temperature Physics. North-Holland, Amsterdam, Chap. 2 (1955)
6. Feynman, R.P.: Statistical Mechanics: A Set of Lectures (Advanced book classics), Chap. 11
7. Haldane, F.D.M., Wu, Y.S.: Quantum dynamics and statistics of vortices in two-dimensional superfluids. Phys. Rev. Lett. **55**, 2887 (1985)
8. Ao, P., Thouless, D.J.: Berry's phase and the magnus force for a vortex line in a superconductor. Phys. Rev. Lett. **70**, 2158 (1993)
9. Stone, M.: Magnus force on skyrmions in ferromagnets and quantum hall systems. Phys. Rev. B **53**, 16573 (1996)
10. Zang, J., Mostovoy, M., Han, J.H., Nagaosa, N.: Dynamics of skyrmion crystals in metallic thin films. Phys. Rev. Lett. **107**, 136804 (2011)
11. Thiele, A.A.: Steady-state motion of magnetic domains. Phys. Rev. Lett. **30**, 230 (1973)
12. Treitiakov, O.A., Clarke, D., Chern, G.-W., Bazaliy, Y.B., Tchernyshyov, O.: Dynamics of domain walls in magnetic nanostrips. Phys. Rev. Lett. **100**, 127204 (2008)
13. Makhfudz, I., Krüger, B., Tchernyshyov, O.: Inertia and chiral edge modes of a skyrmion magnetic bubble. Phys. Rev. Lett. **109**, 217201 (2012)
14. Papanicolaou, N., Tomaras, T.N.: Dynamics of magnetic vortices. Nuc. Phys. B **360**, 425 (1991)
15. Vinen, W.F.: The detection of single quanta of circulation in liquid helium II. Proc. Roy. Soc. A (London) **260**, 218 (1961)
16. Büttner, F., Moutafis, C., Schneider, M., Krüger, B., Günther, C.M., Geilhufe, J., Schmising, C.V.K., Mohanty, J., Pfau, B., Schaffert, S., Bisig, A., Foerster, M., Schulz, T., Vaz, C.A.F., Franken, J.H., Swagten, H.J.M., Kläui, M., Eisebitt, S.: Dynamics and inertia of skyrmionic spin structures. Nat. Phys. **11**, 225 (2015)
17. Mochizuki, M.: Spin-wave modes and their intense excitation effects in skyrmion crystals. Phys. Rev. Lett. **108**, 017601 (2012)
18. Onose, Y., Okamura, Y., Seki, S., Ishiwata, S., Tokura, Y.: Observation of magnetic excitations of skyrmion crystal in a helimagnetic insulator Cu_2OSeO_3. Phys. Rev. Lett. **109**, 037603 (2012)

19. Schwarze, T., Waizner, J., Garst, M., Bauer, A., Stasinopoulos, I., Berger, H., Pfleiderer, C., Grundler, D.: Universal helimagnon and skyrmion excitations in metallic, semiconducting and insulating chiral magnets. Nat. Mater. **14**, 478 (2015)
20. Garst, M., Waizner, J., Grundler, D.: Collective spin excitations of helices and magnetic skyrmions: review and perspectives of magnonics in non-centrosymmetric magnets. arXiv:1702.03668 (2017)

Chapter 5
Skyrmion-Electron Interaction

A typical physical environment in which skyrmions can nucleate is a magnetic metal. In magnetic metals, conduction electrons interact with the spin texture through Hund's rule coupling, which tends to align the spin of the electrons with the orientation of localized moment, creating what is known as the spin transfer torque (STT) effect. As a consequence of the STT mechanism, skyrmions can be driven through the material with an electric current. A modified LLG equation can be derived that includes the presence of STT coupling of electrons with the local moment. If the STT mechanism is describing the action of a conduction electron's spin on the magnetic moment, then an inverse effect, i.e., where the magnetic moment influences the electron dynamics, must also exist. In a manifestation of emergent electrodynamics and as an inverse effect of STT, electrons see the skyrmion spin texture as the source of a magnetic field, and a moving skyrmion as the source of an electric field. Such modifications of the electron dynamics are examined in this chapter for both nonrelativistic and relativistic electrons in two dimensions.

5.1 General Theory of Spin-Transfer Torque

In magnetic metals, the outer-shell electrons of magnetic atoms participate in the formation of the localized moment $\mathbf{n}$, as well as acting as mobile carriers of the electric current. The two types of electrons (i.e., localized and itinerant) originating from the same atom are bound by the Hund's coupling, $-J_{\mathrm{H}}\mathbf{n}\cdot(\Psi^{\dagger}\boldsymbol{\sigma}\Psi)$, where the spinor field Ψ consists of the spin-up and spin-down conduction electron pair

$$\Psi = \begin{pmatrix} c_{\uparrow} \\ c_{\downarrow} \end{pmatrix}. \tag{5.1}$$

J.H. Han, *Skyrmions in Condensed Matter*, Springer Tracts in Modern Physics 278, https://doi.org/10.1007/978-3-319-69246-3_5

Nonrelativistic electrons whose spins are coupled to localized moments in this manner are described by the minimal Hamiltonian

$$H_e = \int d^d\mathbf{r}\, \Psi^\dagger \left(\frac{\mathbf{p}^2}{2m} - J_{\mathrm{H}} \boldsymbol{\sigma} \cdot \mathbf{n} \right) \Psi, \tag{5.2}$$

where the chemical potential of the electrons is omitted for simplicity. This simple model captures the way a flow of electrons, along with their spin flow, can influence the magnetization dynamics. Readers familiar with the spintronics literature will readily recognize this expression as a statement of the spin-transfer torque. A skyrmion, being an example of a nontrivial magnetic texture, is influenced through the STT by the flow of electron spins. In an attempt to derive the modified equation of skyrmion motion due to the electric current, we first survey the general theory of the STT within the Lagrangian framework.

The minimal model Hamiltonian (5.2) can be rephrased as the Lagrangian

$$L_e = i\hbar \Psi^\dagger \partial_t \Psi - \Psi^\dagger \left(\frac{\mathbf{p}^2}{2m} - J_{\mathrm{H}} \boldsymbol{\sigma} \cdot \mathbf{n} \right) \Psi. \tag{5.3}$$

Owing to the ferromagnetic Hund coupling, the spin of the conduction electron tends to align itself with the orientation of the localized moment $\mathbf{n}$ as the electron traverses through space. Metallic systems in which skyrmions have been observed tend to have weak exchange energy, $J \sim 1$ meV, with a corresponding magnetization dynamics time scale $\tau = \hbar/J \sim 10^{-12} - 10^{-13}$s. In contrast, electrons traveling at a speed of $v_f = 10^6$ m/s will spend roughly 10^{-15}-10^{-16}s in the vicinity of each atom. Thus, from the mobile electron's point of view, the local magnetic moments gyrate at a snail's pace, and the electron can align its own spin orientation with the local moment direction within $\hbar/J_{\mathrm{H}} \sim 10^{-15}$s, assuming the modest value $J_{\mathrm{H}} \sim 1$ eV. This is comparable to the time that electrons spend visiting each atom, and therefore, the spin alignment must be more or less complete. As a mathematical idealization, one can consider the $J_{\mathrm{H}} \to \infty$ in order to explore various consequences of strong Hund's rule coupling. In this case, only those electronic degrees of freedom with perfect spin alignment will survive in the physical Hilbert space.

The program of truncating out the unwanted spin orientations in the electronic sector is carried out with the help of spacetime-dependent SU(2) unitary transformations of the spinor field $\Psi = U(\mathbf{r}, t)\psi$, which satisfy

$$U^\dagger(\mathbf{r}, t) \Big[\boldsymbol{\sigma} \cdot \mathbf{n}(\mathbf{r}, t) \Big] U(\mathbf{r}, t) = \sigma^z, \tag{5.4}$$

for an arbitrary spacetime-dependent configuration $\mathbf{n}(\mathbf{r}, t)$. In this basis, the Hund's coupling becomes diagonal in the spin space,

$$\Psi^\dagger (\mathbf{n} \cdot \boldsymbol{\sigma}) \Psi = \psi^\dagger \sigma^z \psi, \quad \psi = \begin{pmatrix} \psi_+ \\ \psi_- \end{pmatrix}, \tag{5.5}$$

and the upper (lower) component spinor ψ_+ (ψ_-) refers to an electron whose spin orientation is along $+\mathbf{n}$ ($-\mathbf{n}$). A specific parametrization of the SU(2) unitary matrix U is offered by

$$U = \mathbf{m}\cdot\boldsymbol{\sigma}, \qquad \mathbf{m} = \left(\sin\frac{\theta}{2}\cos\phi, \sin\frac{\theta}{2}\sin\phi, \cos\frac{\theta}{2}\right), \tag{5.6}$$

for $\mathbf{n} = (\sin\theta\cos\phi, \sin\theta\sin\phi, \cos\theta)$. The electron Lagrangian in the new basis ψ is

$$L_e = \hbar\psi^\dagger\left(i\partial_t + i[U^\dagger\partial_t U]\right)\psi - \psi^\dagger\left(\frac{[\mathbf{p} - i\hbar U^\dagger\nabla U]^2}{2m} - J_{\rm H}\sigma^z\right)\psi. \tag{5.7}$$

The derivatives have turned into covariant derivatives, and are coupled to matrix-valued gauge fields

$$\mathcal{A}_\mu = -iU^\dagger\partial_\mu U = \mathbf{a}_\mu\cdot\boldsymbol{\sigma} = \begin{pmatrix} a_\mu^3 & a_\mu^1 - ia_\mu^2 \\ a_\mu^1 + ia_\mu^2 & -a_\mu^3\end{pmatrix}. \tag{5.8}$$

The electrons see the spin texture $\mathbf{n}$ as a source of a nonabelian gauge field.

The gauge fields $\mathbf{a}_\mu = (a_\mu^1, a_\mu^2, a_\mu^3)$ reflect the underlying spin texture and its dynamics in various ways. First of all, they are related to the $\mathbf{m}$-vector defining the unitary matrix (5.6) by the simple formula

$$\mathbf{a}_\mu = \mathbf{m}\times\partial_\mu\mathbf{m}. \tag{5.9}$$

More explicitly, each component of $\mathbf{a}_\mu = (a_\mu^1, a_\mu^2, a_\mu^3)$ is given by

$$\begin{pmatrix} a_\mu^1 \\ a_\mu^2 \\ a_\mu^3 \end{pmatrix} = \frac{1}{2}\begin{pmatrix} -\sin\phi \\ \cos\phi \\ 0 \end{pmatrix}\partial_\mu\theta + \frac{1}{2}\begin{pmatrix} -\sin\theta\cos\phi \\ -\sin\theta\sin\phi \\ 1-\cos\theta \end{pmatrix}\partial_\mu\phi. \tag{5.10}$$

A clear, geometric meaning can be assigned to the first two components of the gauge field by putting them together as a complex function:

$$a_\mu^1 + ia_\mu^2 = \frac{i}{2}e^{i\phi}(\hat{\theta} + i\hat{\phi})\cdot\partial_\mu\mathbf{n}. \tag{5.11}$$

Here, the two additional unit vectors

$$\begin{aligned} \hat{\theta} &= (\cos\theta\cos\phi, \cos\theta\sin\phi, -\sin\theta) \\ \hat{\phi} &= (-\sin\phi, \cos\phi, 0) \end{aligned} \tag{5.12}$$

form a local orthogonal triad of unit vectors together with $\mathbf{n}$: $\hat{\theta}\times\hat{\phi} = \mathbf{n}$. In this basis, all the "twists" and "turns" of the spin texture $\partial_\mu\mathbf{n}$ show up as $a_\mu^1 + ia_\mu^2$. The interpretation of the far more significant gauge field a_μ^3 will be given in Sect. 5.3, where we examine the phenomenon of emergent electrodynamics in detail. For now, we are still worrying about the influence that electrons exert on the spin dynamics, not the other way around.

With such nonabelian gauge fields in place one can rewrite the Lagrangian as

$$L_e = \hbar\psi^\dagger \left(i\partial_t - \mathcal{A}_0\right)\psi - \psi^\dagger \left(\frac{[\mathbf{p} + \hbar\mathcal{A}]^2}{2m} - J_{\mathrm{H}}\sigma^z\right)\psi. \tag{5.13}$$

The upper component of ψ refers to electrons that faithfully keep their spin orientations aligned with the local moment $\mathbf{n}$ and are rewarded with an energy gain of $2J_{\mathrm{H}}$, compared to the lower spinor component whose spins are always anti-parallel. Off-diagonal elements of the nonabelian gauge field mediate some coupling between the parallel and anti-parallel spins of the electrons. The influence of such off-diagonal terms can be treated in second-order perturbation theory involving the energy gap $2J_{\mathrm{H}}$ in the denominator as such perturbation theories always do. In the infinite-J_{H} limit the perturbative effect vanishes, leaving the two spin orientations completely decoupled. Confined in the truncated Hilbert space of ψ_+ alone, the Lagrangian simplifies to (rewriting ψ_+ as ψ)

$$L_e = \hbar\psi^\dagger[i\partial_t - a_0]\psi - \psi^\dagger\left(\frac{(\mathbf{p} + \hbar\mathbf{a})^2}{2m} + \frac{\hbar^2}{8m}\sum_\mu(\partial_\mu\mathbf{n})^2\right)\psi, \tag{5.14}$$

where $\mathbf{a} = \frac{1}{2}(1 - \cos\theta)\nabla\phi$. In deriving this expression, the relation $\sum_\mu[(a_\mu^1)^2 + (a_\mu^2)^2] = \sum_\mu(\partial_\mu\mathbf{n})^2/4$ was used to arrive at the last term in the Lagrangian. Note also that the Hund energy J_{H} no longer makes an appearance. Equation (5.14) summarizes the interaction of local moments with the conduction electrons in the infinite-J_{H} limit. The procedure we adopted in "bending" the local electron spin orientation to deduce the emergent gauge field a_μ can be found, for instance, in Volovik's 1987 paper [1]. The same technique was applied to deduce the spin-transfer torque in Ref. [2], as we discuss next.

Our goal is to understand the way the electrical current influences the skyrmion dynamics through the effective action (5.14). The strong Hund's coupling enforces the electrons to constantly to readjust their spin orientations to align with the local moment direction. As turning the electron's spin around requires a torque (which must come from the local moment), there must also be a torque acting in the opposite fashion that forces the local moment to rotate the other way. This is the microscopic mechanism underlying the STT. To derive it, we restore the spin dynamics and consider the total Lagrangian

$$L_{e,\mathbf{n}} = -\frac{2S\hbar}{a^d}a_0 + \hbar\psi^\dagger[i\partial_t - a_0]\psi$$
$$-H_{\mathbf{n}} - \psi^\dagger\left(\frac{(\mathbf{p}+\hbar\mathbf{a})^2}{2m} + \frac{\hbar^2}{8m}\sum_\mu(\partial_\mu\mathbf{n})^2\right)\psi. \tag{5.15}$$

Recalling the identity, $\delta a_\mu/\delta\mathbf{n} = (1/2)\partial_\mu\mathbf{n}\times\mathbf{n}$ [cf. (1.66)], variation of the action $\int dt L_{e,\mathbf{n}}$ yields the equation of motion for $\mathbf{n}$:

$$\hbar\left(\frac{S}{a^d} + \frac{1}{2}\psi^\dagger\sigma^z\psi\right)\partial_t\mathbf{n}\times\mathbf{n} + \frac{\delta H'_{\mathbf{n}}}{\delta\mathbf{n}} + \hbar[(\mathbf{j}^e\cdot\nabla)\mathbf{n}]\times\mathbf{n} = 0. \tag{5.16}$$

The modified Hamiltonian $H'_{\mathbf{n}} = H_{\mathbf{n}} + (\hbar^2/8m)\psi^\dagger\psi\sum_\mu(\partial_\mu\mathbf{n})^2$ differs from the bare spin Hamiltonian $H_{\mathbf{n}}$ by a renormalization of the exchange energy $J\to J+\hbar^2\rho_e/4m$, where ρ_e is the electron density. As far as the magnetization dynamics are concerned, one does not need to distinguish $H'_{\mathbf{n}}$ from $H_{\mathbf{n}}$ and J can be treated as a phenomenological constant.

The electronic three-current density $j^e_\mu = (j^e_0, \mathbf{j}^e)$ is then[1]

$$j^e_0 = \psi^\dagger\psi$$
$$\mathbf{j}^e = \frac{1}{2m}\psi^\dagger[\mathbf{p}\psi] - \frac{1}{2m}[\mathbf{p}\psi^\dagger]\psi + \frac{\hbar\mathbf{a}}{m}\psi^\dagger\psi. \tag{5.17}$$

The equation governing the spin dynamics (5.16) can then be cast in a more familiar form by multiplying from the left with $\mathbf{n}\times$,

$$\hbar\left(\frac{S}{a^d} + \frac{1}{2}\rho_e\right)\partial_t\mathbf{n} = \frac{\delta H_{\mathbf{n}}}{\delta\mathbf{n}}\times\mathbf{n} - \hbar(\mathbf{j}^e\cdot\nabla)\mathbf{n}. \tag{5.18}$$

This is the modified Landau–Lifshitz equation governing spin dynamics in the presence of STT coupling to the electric current. The spin density S/a^d of the purely magnetic theory is now modified to encompass the overall spin density as $S/a^d + \rho_e/2 = (S+n_e/2)/a^d$, where n_e is the filling factor of the conduction electrons.

If one takes $\delta H_{\mathbf{n}}/\delta\mathbf{n}\times\mathbf{n} = 0$ and assumes $\mathbf{n}(\mathbf{r},t) = \mathbf{n}(\mathbf{r}-\mathbf{v}_s t)$, i.e., a uniform drift of the ground state spin configuration, the drift velocity of the spin $\mathbf{v}_s$ is related to that of the electron $\mathbf{v}_e$ (arising from $\mathbf{j}^e = (n_e/a^d)\mathbf{v}_e$), through

$$\left(S + \frac{n_e}{2}\right)\mathbf{v}_s = n_e\mathbf{v}_e. \tag{5.19}$$

[1] Note that the electric charge does not enter in the definition of current. To be precise, j^e_μ is a three-current for the electron number density. To clarify this difference, we refer to j^e_μ as the "electronic current" from time to time instead of the familiar "electric current".

5.1.1 SU(2) Rotation on a Lattice

The kinetic energy $\mathbf{p}^2/2m$ picks up the gauge term and becomes $(\mathbf{p}+\hbar\mathbf{a})^2/2m$ under the SU(2) rotation of the spinor $\Psi = U\psi$. Although the emergent vector potential $\mathbf{a}$ itself is not gauge invariant, its curl $\nabla \times \mathbf{a}$ represented a physical quantity related to the underlying spin texture. It is worth seeing how the SU(2) rotation plays out in the lattice model.

With the spinor Ψ_i defined at each lattice site i, the SU(2) gauge transformation leads to the altered hopping

$$\Psi_j^\dagger \Psi_i \to \psi_j^\dagger U_j^\dagger U_i \psi_i, \tag{5.20}$$

where each $U_i = \mathbf{m}_i \cdot \boldsymbol{\sigma}$ implements the rotation

$$\psi_i^\dagger U_i^\dagger (\mathbf{n}_i \cdot \boldsymbol{\sigma}) U_i \psi = \psi_i^\dagger \sigma^z \psi_i, \tag{5.21}$$

using the same $\mathbf{m}_i$ vector given earlier in (5.6). After some algebraic manipulation one can show

$$U_j^\dagger U_i = (\mathbf{m}_j \cdot \boldsymbol{\sigma})(\mathbf{m}_i \cdot \boldsymbol{\sigma}) = \mathbf{m}_j \cdot \mathbf{m}_i + i(\mathbf{m}_j \times \mathbf{m}_i) \cdot \boldsymbol{\sigma}. \tag{5.22}$$

The hopping matrix element is generally complex, and the parallel component of the electron ψ_{i+} is seen to hop with an amplitude

$$t_{ji} \equiv \mathbf{m}_j \cdot \mathbf{m}_i + i(\mathbf{m}_j \times \mathbf{m}_i)_z = \mathbf{m}_j \cdot \mathbf{m}_i + i(\mathbf{m}_j \times \mathbf{m}_i) \cdot \mathbf{n}_0, \tag{5.23}$$

where $\mathbf{n}_0 = (0, 0, 1)$. The amplitude of the hopping works out to be

$$|t_{ji}|^2 = (\mathbf{m}_j \cdot \mathbf{m}_i)^2 + (\mathbf{m}_j \times \mathbf{m}_i \cdot \mathbf{n}_0)^2 = \frac{1}{2}(1 + \mathbf{n}_j \cdot \mathbf{n}_i), \tag{5.24}$$

which is reduced from 1 for spins that are not in perfect parallel alignment.

Imagine a parallel-spin electron ψ_+ hopping consecutively along the three sites $i \to j$, $j \to k$, and back to i. The product of three hopping parameters around a triangle $i \to j \to k \to i$ is

$$\begin{aligned} t_{ik}t_{kj}t_{ji} &= [\mathbf{m}_i \cdot \mathbf{m}_k + i(\mathbf{m}_i \times \mathbf{m}_k) \cdot \mathbf{n}_0][\mathbf{m}_k \cdot \mathbf{m}_j + i(\mathbf{m}_k \times \mathbf{m}_j) \cdot \mathbf{n}_0] \\ &\quad \times [\mathbf{m}_j \cdot \mathbf{m}_i + i(\mathbf{m}_j \times \mathbf{m}_i) \cdot \mathbf{n}_0] \\ &= \frac{1}{4}\left[1 + \mathbf{n}_i \cdot \mathbf{n}_j + \mathbf{n}_j \cdot \mathbf{n}_k + \mathbf{n}_k \cdot \mathbf{n}_i - i(\mathbf{n}_i \cdot \mathbf{n}_j \times \mathbf{n}_k)\right] \\ &= \sqrt{\frac{1 + \mathbf{n}_i \cdot \mathbf{n}_j}{2}}\sqrt{\frac{1 + \mathbf{n}_j \cdot \mathbf{n}_k}{2}}\sqrt{\frac{1 + \mathbf{n}_k \cdot \mathbf{n}_i}{2}}\, e^{-i\Phi_{ijk}/2}. \end{aligned} \tag{5.25}$$

The final expression for the three-hop amplitude is expressed solely in terms of the $\mathbf{n}$-vectors alone (not involving $\mathbf{n}_0$), signaling the gauge-invariance of the quantity. The phase factor Φ_{ijk} is given through the definition

$$\tan\left[\frac{\Phi_{ijk}}{2}\right] = \frac{\mathbf{n}_i \cdot \mathbf{n}_j \times \mathbf{n}_k}{1 + \mathbf{n}_i \cdot \mathbf{n}_j + \mathbf{n}_j \cdot \mathbf{n}_k + \mathbf{n}_k \cdot \mathbf{n}_i}. \tag{5.26}$$

This is the lattice generalization of the spin chirality, which is usually defined as the triple product of three spins $\mathbf{n}_i \cdot (\mathbf{n}_j \times \mathbf{n}_k)$. Furthermore, we find that electrons moving around a closed loop $\langle ijk \rangle$ accumulates an Aharonov-Bohm phase factor $\Phi_{ijk}/2$ due to the emergent flux. In fact, Φ_{ijk} is the solid angle subtended by the three spins $(\mathbf{n}_i, \mathbf{n}_j, \mathbf{n}_k)$ on the unit sphere. An electron making a full round of the skyrmion would pick up the phase $\sum \Phi_{ijk}/2 = 4\pi/2 = 2\pi$, just as in going round a flux quantum.

5.2 Skyrmion Dynamics Under the Spin-Transfer Torque

In Sect. 4.2.2, the LLG equation was used to derive the equation of motion governing skyrmion dynamics. In the previous section we have derived the modified LLG equation (5.18) to describe the general spin dynamics in the presence of the spin-transfer torque. Now we try to derive the updated form of the skyrmion dynamics, starting from the modified LLG equation.

First, we recall that in Chap. 4 a rigid skyrmion ansatz $\mathbf{n}(\mathbf{r}, t) = \mathbf{n}_0(\mathbf{r} - \mathbf{R}(t))$ was used to derive the point-particle skyrmion dynamics. We also saw that variations in the spin action were entirely due to the displacement $\delta\mathbf{R} = (\delta X, \delta Y)$ of the skyrmion coordinates. Most importantly, the variation of the emergent gauge field a_μ was related to $\delta\mathbf{R}$ through

$$\begin{aligned} \delta a_\mu &= \frac{1}{2}\delta\mathbf{n} \cdot [\partial_\mu \mathbf{n} \times \mathbf{n}] = -\frac{1}{2}(\delta X \partial_x \mathbf{n} + \delta Y \partial_y \mathbf{n}) \cdot [\partial_\mu \mathbf{n} \times \mathbf{n}] \\ (\delta a_x, \delta a_y, \delta a_0) &= 2\pi q_s(\delta Y, -\delta X, \delta X\, \dot{Y} - \delta Y\, \dot{X}). \end{aligned} \tag{5.27}$$

The end result of our previous analysis was the skyrmion's point-particle Lagrangian (4.61), which we reproduce below:

$$L_s = \frac{M_s}{2}\dot{\mathbf{R}}^2 - \frac{G}{2}\hat{z} \cdot (\mathbf{R} \times \dot{\mathbf{R}}) - V(\mathbf{R}). \tag{5.28}$$

The expression for the electron-spin Lagrangian (5.15) makes it clear that the electron-spin coupling takes place via emergent gauge fields. Variation of the temporal coupling part of the action gives

$$\delta\Big(\hbar\int d^2\mathbf{r}dt\ \psi^\dagger[-a_0]\psi\Big) = -h\Big(\int d^2\mathbf{r}\ q(\mathbf{r})j_0^e(\mathbf{r})\Big)\int dt(\delta X\ \dot{Y} - \delta Y\ \dot{X})$$
$$= -hJ_0^e\int dt(\delta X\ \dot{Y} - \delta Y\ \dot{X}). \tag{5.29}$$

The weighted electronic current is defined as $J_\mu^e = \int d^2\mathbf{r}\ q(\mathbf{r})j_\mu^e(\mathbf{r})$ (compare with the definition of the electronic current density j_μ^e in (5.17)). Similarly, variation of the spatial part of the action yields

$$\delta\Big(-\frac{1}{2m}\int d^2\mathbf{r}dt\ \psi^\dagger[\mathbf{p}+\hbar\mathbf{a}]^2\psi\Big) = -\hbar\int d^2\mathbf{r}dt\ \delta\mathbf{a}\cdot\mathbf{j}^e$$
$$= h\int dt\ (\delta X\ J_y^e - \delta Y\ J_x^e). \tag{5.30}$$

Both variations (5.29) and (5.30) are reproduced by the Lagrangian

$$L_s = -\frac{1}{2}hJ_0^e\hat{z}\cdot(\mathbf{R}\times\dot{\mathbf{R}}) + h\hat{z}\cdot(\mathbf{R}\times\mathbf{J}^e). \tag{5.31}$$

The overall Lagrangian for a single skyrmion subject to STT coupling is

$$L_s = \frac{1}{2}M_s\dot{\mathbf{R}}^2 - \frac{1}{2}[G + hJ_0^e]\hat{z}\cdot(\mathbf{R}\times\dot{\mathbf{R}}) + h\hat{z}\cdot(\mathbf{R}\times\mathbf{J}^e) - V(\mathbf{R}). \tag{5.32}$$

The modified skyrmion equation of motion, including the Gilbert damping effect, reads

$$M_s\ddot{\mathbf{R}} = -\frac{\partial V}{\partial\mathbf{R}} + \hat{z}\times\Big(\big[G + hJ_0^e\big]\dot{\mathbf{R}} - h\mathbf{J}^e\Big) + \frac{\alpha\eta}{Q_s}G\dot{\mathbf{R}}. \tag{5.33}$$

For a massless skyrmion in a homogeneous background, $V(\mathbf{R}) = 0$, the skyrmion drift velocity assumes the simple result

$$\dot{\mathbf{R}} \simeq \frac{h\mathbf{J}^e}{G+hJ_0^e} - \frac{\alpha\eta}{Q_s}\frac{G}{G+hJ_0^e}\hat{z}\times\left(\frac{h\mathbf{J}^e}{G+hJ_0^e}\right). \tag{5.34}$$

In the absence of damping, the drift velocity of the skyrmion is related to the electron current by $\mathbf{v}_s = h\mathbf{J}^e/(G + hJ_0^e)$. Using this definition of $\mathbf{v}_s$, one can rewrite the original drift velocity formula (5.34) as

$$\dot{\mathbf{R}} \simeq \mathbf{v}_s - \frac{\alpha\eta}{Q_s}\frac{G}{G+hJ_0^e}\hat{z}\times\mathbf{v}_s. \tag{5.35}$$

Interestingly, the transverse motion of the skyrmion is caused by the dissipative effect $\sim\alpha$. For the electrons, the transverse motion is induced by the nondissipative Lorentz force.

5.3 Nonrelativistic Fermions Coupled to a Skyrmion Texture

In Sect. 5.1, we considered a minimal model Lagrangian for nonrelativistic electrons that were strongly Hund-coupled to a localized spin texture. We reproduce here the electron Lagrangian obtained after the unitary rotation to the local magnetization direction [cf. (5.14)]

$$L_e = \hbar\psi^\dagger[i\partial_t - a_0]\psi - \psi^\dagger\left(\frac{(\mathbf{p}+\hbar\mathbf{a})^2}{2m} + \frac{\hbar^2}{8m}\sum_\mu(\partial_\mu\mathbf{n})^2\right)\psi.$$

The emergent gauge fields a_μ, derived from the underlying spin texture, assumes the same role in the Lagrangian as the physical electromagnetic gauge fields. By direct comparison of the emergent gauge field to electromagnetic field coupling, we can see the clear correspondence

$$\begin{aligned} \hbar a_0 &\leftrightarrow q_e A_0 \\ \hbar\mathbf{a} &\leftrightarrow -q_e\mathbf{A} \end{aligned} \tag{5.36}$$

between the emergent $(a_0, \mathbf{a})$ and electromagnetic $(A_0, \mathbf{A})$ gauge fields. This correspondence extends to the emergent electric and magnetic fields

$$\begin{aligned} \mathbf{b} &= -\frac{\hbar}{q_e}\nabla\times\mathbf{a} \\ \mathbf{e} &= -\frac{\hbar}{q_e}(\nabla a_0 - \partial_t\mathbf{a}), \end{aligned} \tag{5.37}$$

which have exactly the same effect on the electron dynamics as electric and magnetic fields of electromagnetic origin. Both $(\mathbf{b}, \mathbf{e})$ fields can be combined to form a field tensor

$$f_{\mu\nu} = -\frac{\hbar}{q_e}(\partial_\mu a_\nu - \partial_\nu a_\mu) = -\frac{\hbar}{2q_e}\mathbf{n}\cdot(\partial_\mu\mathbf{n}\times\partial_\nu\mathbf{n}). \tag{5.38}$$

In terms of the spin texture, the emergent electromagnetic fields are

$$\begin{aligned} -\frac{\hbar}{2q_e}\mathbf{n}\cdot(\partial_\mu\times\partial_\nu\mathbf{n}) &= \varepsilon_{\mu\nu\lambda}b_\lambda, \\ -\frac{\hbar}{2q_e}\mathbf{n}\cdot(\partial_\lambda\mathbf{n}\times\partial_t\mathbf{n}) &= e_\lambda. \end{aligned} \tag{5.39}$$

According to the above prescription, an electron sees one skyrmion of topological charge Q_s as the magnetic flux

$$\Phi = -\frac{\hbar}{q_e}\int dxdy\, b_z = -\frac{h}{q_e}Q_s. \tag{5.40}$$

Each skyrmion with unit charge is seen as a flux quantum to the electron. One skyrmion charge confined to the 10 nm×10 nm area is equivalent to the magnetic field of magnitude

$$|b_z| \times 10^{-16}\text{m}^2 = 4.134 \times 10^{-15}\text{Wb} \to b \approx 41\text{T}! \tag{5.41}$$

This is an enormous magnetic field, hard to achieve even for super-modern high magnetic field facilities. The emergent skyrmionic magnetic fields thus open up the possibility of observing strong Hall effects. Reflective of its topological character, the skyrmion-induced Hall effect by the electrons is also known as the topological Hall effect. In fact, the observation of the topological Hall effect was one of the earliest indicators of the formation of a nontrivial spin texture in a metallic ferromagnet. In 2009, a significant downward deviation from the classical Hall signal, $\sigma_{xy} \propto B$, where B is the external magnetic field, was reported over a certain magnetic field range [3, 4], and was eventually interpreted as signaling the formation of skyrmions carrying the emergent flux opposite to the external field.

A moving spin texture, which acts like a moving flux in electrodynamics, can be viewed as a source of electric field, in accordance with Faraday's law. In the field of spintronics, this extra electric field acting on the conduction electrons arising from the moving spin texture is called the spin-motive force [5]. It was shown in Sect. 5.2 that a uniform electronic current induces the drift motion of the skyrmion through the spin-transfer torque mechanism. The uniform skyrmion drift, in turn, allowed one to write $\mathbf{n}(\mathbf{r}, t) = \mathbf{n}(\mathbf{r} - \mathbf{v}_s t)$. Replacing $\partial_t \mathbf{n}$ in (5.39) by $-(\mathbf{v}_s \cdot \nabla)\mathbf{n}$, the emergent electric field $\mathbf{e}$ due to a moving skyrmion can be written as

$$\mathbf{e} = -\frac{h}{q_e}\rho_s \hat{z} \times \mathbf{v}_s = \mathbf{b} \times \mathbf{v}_s, \tag{5.42}$$

with the skyrmion density $\rho_s = \mathbf{n} \cdot (\partial_x \mathbf{n} \times \partial_y \mathbf{n})/4\pi$ and the emergent magnetic field $\mathbf{b} = b_z \hat{z}$ in place.

It is a good exercise to walk carefully through the various electromagnetic and emergent forces that act on the electrons. According to the Lorentz's force law $q_e \mathbf{v}_e \times \mathbf{B}$, an external magnetic field $\mathbf{B} = B\hat{z}$ pointing out of the page tends to deflect a right-moving electron with initial velocity $\mathbf{v}_e = v_e \hat{x}$ upwards, i.e., in the $+\hat{y}$ direction. On the other hand, the emergent magnetic field created by the skyrmion is much larger than a typical external field, and has the opposite sign. As a result, the net deflection of the electron is downward, $-\hat{y}$, as shown in Fig. 5.1. At the same time, a uniform current flow causes the skyrmion to drift parallel to the electron drift with velocity $\mathbf{v}_s$, generating an emergent electric field $\mathbf{e} = \mathbf{b} \times \mathbf{v}_s$ which points downward, $\mathbf{e} \propto -\hat{y}$. The resulting spin-motive force $q_e \mathbf{e}$ points up, $+\hat{y}$, thus lessening the amount of deflection created by the emergent magnetic field of a static skyrmion. We conclude

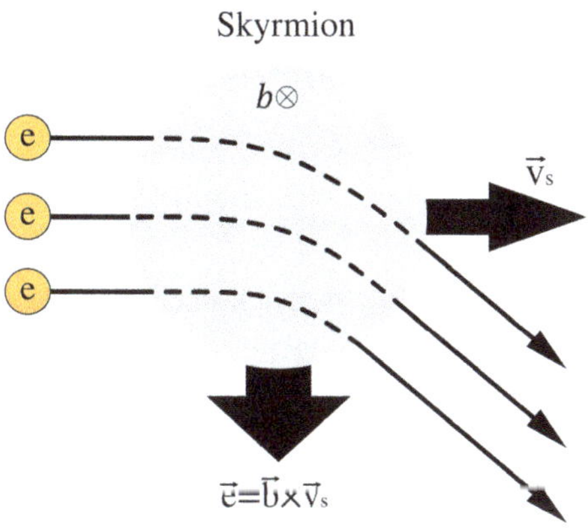

Fig. 5.1 Electrons see the skyrmion as a source of flux and get deflected as they pass through the skyrmion on account of strong Hund's rule coupling. A uniform electron drift leads to a corresponding skyrmion drift motion with velocity $\mathbf{v}_s$. Seeing the moving skyrmion as a source of electric field $\mathbf{e}$ through Faraday's law, electrons feel an additional transverse force which acts opposite to the emergent Lorentz force

that the Hall effect arising from a mobile skyrmion is less than that arising from a stationary one.

Such heuristic discussion can be made more formal using the Drude theory of electron motion encompassing both electromagnetic and emergent fields:

$$m\frac{d\mathbf{v}_e}{dt} = q_e(\mathbf{e}+\mathbf{E}) + q_e\mathbf{v}_e \times (\mathbf{b}+\mathbf{B}) - \frac{m\mathbf{v}_e}{\tau}$$
$$= q_e\mathbf{E} + q_e(\mathbf{v}_e - \mathbf{v}_s)\times\mathbf{b} + q_e\mathbf{v}_e\times\mathbf{B} - \frac{m\mathbf{v}_e}{\tau}. \tag{5.43}$$

The emergent Lorentz force acting on the electron is governed by the relative velocity of the electron to the skyrmion's drift velocity, $\mathbf{v}_e - \mathbf{v}_s$, as expected from Galilean invariance. When the skyrmions are pinned so that $\mathbf{v}_s = 0$, the electron drift velocity $\mathbf{v}_{e,\parallel}$ along the $\mathbf{E}$-field is obtained, to first order, by $\mathbf{v}_{e,\parallel} = (q_e\tau/m)\mathbf{E}$. The transverse component of the velocity follows from the perturbative calculation

$$\mathbf{v}_{e,\perp} \simeq \frac{q_e\tau}{m}\mathbf{v}_{e,\parallel}\times(\mathbf{b}+\mathbf{B}) = \left(\frac{q_e\tau}{m}\right)^2 \mathbf{E}\times(\mathbf{b}+\mathbf{B}). \tag{5.44}$$

The Hall current $\mathbf{j}_\mathrm{H} = q_e n_e \mathbf{v}_{e,\perp}$ is then

$$\mathbf{j}_\mathrm{H} = \sigma_{xx}\frac{q_e\tau}{m}(b+B)\mathbf{E}\times\hat{z}, \qquad (\mathbf{v}_s = 0) \tag{5.45}$$

where we have used the Drude expression for the conductivity $\sigma_{xx} = n_e q_e^2/m$ and set $\mathbf{b} = b\hat{z}$, $\mathbf{B} = B\hat{z}$.

Once the skyrmions are set in motion, $\mathbf{v}_s$ becomes nonzero and assumes a value proportional to the electron drift velocity, $\mathbf{v}_s = f\mathbf{v}_e$, with a phenomenological constant f. The modified Hall current is

$$\mathbf{j}_{\mathrm{H}} = \sigma_{xx}\frac{q_e\tau}{m}\big(b[1-f]+B\big)\mathbf{E}\times\hat{z}. \quad (\mathbf{v}_s \neq 0) \tag{5.46}$$

In practice, an applied electric field sets up a Hall current in accordance with (5.45). For strong enough applied field exceeding the de-pinning threshold, skyrmions previously pinned by impurities are de-pinned and assume steady-state motion with velocity $\mathbf{v}_s$. Consequently, the Hall current drops suddenly to the value given by (5.46). This sudden drop in the Hall conductivity can be used as a sharp indicator of the de-pinning transition of the skyrmion lattice, as demonstrated beautifully in the experiment by the Pfleiderer group [6].

5.4 Relativistic Fermions Coupled to a Skyrmion Texture

Two-dimensional relativistic fermions appear in many contexts of condensed matter such as in two-dimensional graphene materials and the surface of topological insulators. The topological insulator, for instance, is a host to a surface state that carries an odd number of Dirac fermions. For one such Dirac species Ψ, a minimal Hamiltonian can be given as

$$H = \Psi^\dagger\Big(v[(\mathbf{p}-q_e\mathbf{A})\times\boldsymbol{\sigma}]\cdot\hat{z} - J_{\mathrm{H}}\boldsymbol{\sigma}\cdot\mathbf{n}\Big)\Psi. \tag{5.47}$$

The second term, which is the familiar Hund coupling, is added to address the physics of Dirac electrons coupled to the localized moment $\mathbf{n}$. We follow Ref. [7] to discuss the general effect of Hund coupling on the Dirac electrons.

In Sect. 5.3 we discussed at length how to treat the effects of Hund coupling through adiabatic rotation of the electron spin orientation along that of $\mathbf{n}$. Here, we do not need to implement the rotation at all. The above Hamiltonian can be rewritten in the form

$$\begin{aligned} H &= v\Psi^\dagger\Big[\Big(\mathbf{p}-q_e\mathbf{A}+\frac{J_{\mathrm{H}}}{v}\hat{z}\times\mathbf{n}\Big)\times\boldsymbol{\sigma}\cdot\hat{z}\Big]\Psi - J_{\mathrm{H}}n^z\Psi^\dagger\sigma^z\Psi \\ &= v\Psi\Big[(\mathbf{p}-q_e(\mathbf{A}+\mathbf{a}))\times\boldsymbol{\sigma}\cdot\hat{z}\Big]\Psi - J_{\mathrm{H}}n^z\Psi^\dagger\sigma^z\Psi, \end{aligned} \tag{5.48}$$

where the emergent vector potential $\mathbf{a}$ is related to the local magnetization $\mathbf{n}$ by

$$\mathbf{a} = \frac{J_{\mathrm{H}}}{q_e v}\mathbf{n}\times\hat{z}. \text{ (relativistic)} \tag{5.49}$$

This vector potential is very different from the one we had in the nonrelativistic electron problem, which had the gauge-dependent vector potential

$$\mathbf{a} = \frac{\hbar}{2q_e}(\cos\theta - 1)\nabla\phi. \text{ (nonrelativistic)} \tag{5.50}$$

In the relativistic case, it turns out, we do not even need to make the spin rotation to arrive at the emergent vector potential of (5.49). The gauge potential **a** already depends on the physical quantity **n** and remains gauge-invariant. Emergent electromagnetic fields arising from **a** are also quite different from the nonrelativistic examples worked out in the previous section:

$$\begin{aligned} b = \nabla_2 \times \mathbf{a} &= -\frac{J_{\mathrm{H}}}{q_e v}(\nabla_2 \cdot \mathbf{n}) \\ \mathbf{e} = -\partial_t \mathbf{a} &= -\frac{J_{\mathrm{H}}}{q_e v}\dot{\mathbf{n}} \times \hat{z}. \end{aligned} \tag{5.51}$$

One may ask: why did we not make the spin rotation? First of all, there was no need to do so since the Hund coupling was already in a form compatible with the gauge coupling to the Dirac electron. Secondly, the spin rotation method of the previous section ultimately had to rely on the large-J_{H} limit to truncate out the oppositely oriented electron spins. No such approximation is required to justify the vector potential **a** obtained in the relativistic case.

Suppose we have the single skyrmion spin texture $\mathbf{n}_s$. It is immediately clear that the integrated emergent flux is zero, $\int dxdy\, b \propto \int dxdy\, (\nabla_2 \cdot \mathbf{n}_s) = 0$, in stark contrast to a nonrelativistic fermion that sees the skyrmion as one flux quantum. For a similar reason, a moving skyrmion is not equivalent to a moving flux that generates a transverse electric field according to Faraday's law. Substitution of $\mathbf{n}_s(\mathbf{r}, t) = \mathbf{n}_s(\mathbf{r} - \mathbf{R}_s(t))$ gives

$$\mathbf{e} = \frac{J_{\mathrm{H}}}{q_e v}[(\dot{\mathbf{R}} \cdot \nabla)\mathbf{n}_s] \times \hat{z} = \frac{J_{\mathrm{H}}}{q_e v}((\mathbf{v}_s \cdot \nabla)n_y, -(\mathbf{v}_s \cdot \nabla)n_x). \tag{5.52}$$

Faraday relation (5.42) does not apply here, as may be shown clearly by considering the case where the skyrmion velocity is $\mathbf{v}_s = v_s\hat{x}$, in which case both components of **e** are nonzero, while Faraday's relation only predicts a transverse electric field. In any case, the integral of **e** over space vanishes. It appears that the influence of a skyrmion spin texture on a relativistic electron is not as dramatic as it was for nonrelativistic electrons.

On the other hand, a close inspection of (5.48) suggests that the z-component of the local magnetization texture n^z acts like a position-dependent mass for the Dirac electrons. From the well-known Jackiw–Rebbi mechanism, we are accustomed to seeing a fermionic bound state in the region where the mass of a relativistic electron undergoes a sign change. Indeed, a careful calculation of the electronic spectrum, assuming a skyrmion spin texture and Hund coupling, found some localized states in the vicinity of the $n^z = 0$ region [8]. Due to the fermionic bound state, the skyrmion can become electrically charged.

For comparison, the coupling of Dirac electrons to a spin vortex $\mathbf{n}_v = (\cos\phi, \sin\phi, 0)$, where $\phi = \tan^{-1}(y/x) + \phi_0$ has more dramatic effect. The divergence $\nabla_2 \cdot \mathbf{n}_v = \cos\phi_0/r$ here gives rise to the total flux of

$$\int_{r\leq R} dxdy\, b = -2\pi \frac{J_{\rm H}}{q_e v} R\cos\phi_0 \tag{5.53}$$

in a circular region of radius R. A spin vortex then acts as a source of (potentially very large) magnetic fields for Dirac electrons that are Hund-coupled to it. One should keep in mind, however, that a vortex must always be accompanied by an anti-vortex, which creates an opposing magnetic flux for the Dirac electrons.

References

1. Volovik, G.E.: Linear momentum in ferromagnets. J. Phys. C. Solid State Phys. **20**, L83 (1987)
2. Bazaliy, Y.B., Jones, B.A., Zhang, S.C.: Modification of the Landau-Lifshitz equation in the presence of a spin-polarized current in colossal- and giant-magnetoresistive materials. Phys. Rev. B **57**, R3213(R) (1998)
3. Lee, M., Kang, W., Onose, Y., Tokura, Y., Ong, N.P.: Unusual Hall effect anomaly in MnSi under pressure. Phys. Rev. Lett. **102**, 186601 (2009)
4. Neubauer, A., Pfleiderer, C., Binz, B., Rosch, A., Ritz, R., Niklowitz, P.G., Böni, P.: Topological Hall effect in the a phase of MnSi. Phys. Rev. Lett. **102**, 186602 (2009)
5. Barnes, S.E., Maekawa, S.: Generalization of Faraday's law to include nonconservative spin forces. Phys. Rev. Lett. **98**, 246601 (2007)
6. Schulz, T., Ritz, R., Bauer, A., Halder, M., Wagner, M., Franz, C., Pfleiderer, C., Everschor, K., Garst, M., Rosch, A.: Emergent electrodynamics of skyrmions in a chiral magnet. Nat. Phys. **8**, 301 (2012)
7. Nomura, K., Nagaosa, N.: Electric charging of magnetic textures on the surface of a topological insulator. Phys. Rev. B **82**, 161401(R) (2010)
8. Hurst, H.M., Efkimkin, D.K., Zang, J., Galitski, V.: Charged skyrmions on the surface of a topological insulator. Phys. Rev. B **91**, 060401(R) (2015)

Chapter 6
Magnon Dynamics

Small fluctuations of the ordered moments in magnets are known as magnons.[1] For a single skyrmion, geometrically confined spin excitations arise corresponding to various deformations of the ground-state skyrmion texture. The zero-point fluctuations of the confined magnon modes are the origin of the effective mass of the skyrmion. For a lattice of skyrmions, magnons form bands with nontrivial topological Chern numbers. The skyrmions act as a source of emergent flux for the magnons and scatter them preferentially in a particular direction, leading to an observable topological magnon Hall effect. A magnon wave can be tailored to drive the skyrmions, in a manifestation of the magnonic version of the spin-transfer torque. A rich variety of magnon-related phenomena can be understood from the general formulation of the Holstein-Primakoff (HP) theory for an arbitrary spin-textured ground state.

6.1 General Formulation

A general procedure exists for deriving the magnon dynamics in any smooth spin background. The HP transformation is one such technique that is applicable to ferromagnets, antiferromagnets, and spiral magnets. The situation one encounters in the skyrmion problem is considerably more complex, since the ground state spin configuration is textured (twisted) in all directions.

As we learned in Chap. 3, an appropriate continuum Hamiltonian of a chiral ferromagnet governing skyrmion formation consists of Heisenberg, Dzyaloshinskii-Moriya, and Zeeman terms:

[1]Alternatively, spin waves. The two terms are used interchangeably in this chapter.

J.H. Han, *Skyrmions in Condensed Matter*, Springer Tracts
in Modern Physics 278, https://doi.org/10.1007/978-3-319-69246-3_6

$$\begin{aligned} H_{\text{HDMZ}} &= \frac{J}{2}\sum_{\mu=1}^{3}(\partial_\mu \mathbf{n})^2 + D\mathbf{n}\cdot(\nabla\times\mathbf{n}) - JS\mathbf{B}\cdot\mathbf{n} \\ &\Rightarrow \frac{J}{2}\sum_{\mu=1}^{3}[\partial_\mu \mathbf{n} - \kappa\hat{e}_\mu \times \mathbf{n}]^2 - JS\mathbf{B}\cdot\mathbf{n}. \end{aligned} \tag{6.1}$$

The second line is an exact reformulation of the first, up to an irrelevant constant, and is written in terms of three orthogonal Cartesian unit vectors $\hat{e}_\mu$ ($\mu = 1, 2, 3$), and the dimensionless ratio $\kappa = D/J$, which measures the relative strength of the DM and exchange interactions. For later convenience, the magnitude of the classical spin S and the overall energy scale J is explicitly factored out of the Zeeman term. This model can be easily reduced to a two-dimensional model by truncating all $\mu = 3$ terms. As we have seen, possible ground states of this Hamiltonian include the spin spiral, skyrmion crystal, and the hedgehog-anti-hedgehog crystal. The HP theory of small spin fluctuations (i.e., magnons) can be developed around a particular ground state spin configuration denoted by $\mathbf{n}_0(\mathbf{r})$. The only requirement is that $\mathbf{n}_0$ is a slowly-varying function with respect to the atomic lattice spacing. Skyrmions are an ideal problem for such an approximation due to their large size set by $\kappa^{-1} = J/D \gg 1$.

The way to tackle the magnon problem for a general ground state $\mathbf{n}_0$ is to first perform a rotation to a local frame of reference. In the new reference frame, the local spin direction $\mathbf{n}_0$ is aligned with the $\hat{z}$-axis as in a simple ferromagnet, where the application of the Holstein-Primakoff formalism is most straightforward. Readers will recall that an analogous idea was employed when we handled the problem of Hund coupling of electron spin to the local moment in Chap. 5. There, the electron spinor was rotated using an SU(2) matrix. This time, the real-valued spin vector $\mathbf{n}$ is rotated by the SO(3) matrix.

Denoting the ground state spin configuration $\mathbf{n}_0 = (\sin\theta\cos\phi,\ \sin\theta\sin\phi,\ \cos\theta)$ with position-dependent $(\theta,\ \phi)$, the required orthogonal matrix $\mathscr{R}$ is

$$\mathscr{R}_{\alpha\beta} = 2m_\alpha m_\beta - \delta_{\alpha\beta}, \quad \mathbf{m} = \left(\sin\frac{\theta}{2}\cos\phi,\ \sin\frac{\theta}{2}\sin\phi,\ \cos\frac{\theta}{2}\right). \tag{6.2}$$

Once the ground state $\mathbf{n}_0$ is known, there is no ambiguity in defining $\mathbf{m}$, and henceforth $\mathscr{R}$. From the symmetric property of the matrix $\mathscr{R}^T = \mathscr{R} = \mathscr{R}^{-1}$, it can be easily shown that

$$\mathscr{R}\,\hat{z} = \mathbf{n}_0 \ \text{ and } \ \hat{z} = \mathscr{R}\,\mathbf{n}_0. \tag{6.3}$$

Having found the proper rotation matrix, all the $\mathbf{n}$ vectors in the HDMZ Hamiltonian are replaced by $\mathscr{R}\mathbf{n}$, and all $\partial_\mu \mathbf{n}$ by

$$\partial_\mu \mathbf{n} \to \partial_\mu(\mathscr{R}\mathbf{n}) = \mathscr{R}[\,\partial_\mu \mathbf{n} + (\mathscr{R}\partial_\mu \mathscr{R})\mathbf{n}\,]. \tag{6.4}$$

The additional $(\mathscr{R}\partial_\mu \mathscr{R})\mathbf{n}$ term can be reorganized as a vector potential of some sort,

$$
\begin{aligned}
(\mathscr{R}\partial_\mu \mathscr{R})\mathbf{n} &= -\mathbf{a}_\mu \times \mathbf{n}, \\
\mathbf{a}_\mu &= 2\mathbf{m} \times \partial_\mu \mathbf{m} \\
&= \begin{pmatrix} -\sin\phi \\ \cos\phi \\ 0 \end{pmatrix} \partial_\mu \theta + \begin{pmatrix} -\sin\theta\cos\phi \\ -\sin\theta\sin\phi \\ 1-\cos\theta \end{pmatrix} \partial_\mu \phi.
\end{aligned} \tag{6.5}
$$

This vector potential $\mathbf{a}_\mu$ is, interestingly, twice the emergent nonabelian gauge field found for the SU(2) rotation of the electron spinor in Chap. 5 [cf. (5.10)]. This suggests that the same gauge field generated by the skyrmion background will influence the magnon dynamics with twice as much impact as they do the electron dynamics. The detailed calculation that follows proves this assertion to be the case.

The full HDMZ Hamiltonian in the rotated frame is

$$
\frac{H_{\text{HDMZ}}}{J} = \frac{1}{2}\sum_{\mu=1}^{3}\Big(\partial_\mu \mathbf{n} - \mathbf{a}_\mu \times \mathbf{n} - \kappa \mathscr{R}[\hat{e}_\mu \times \mathbf{n}]\Big)^2 - SB\mathbf{n}_0 \cdot \mathbf{n}. \tag{6.6}
$$

The Zeeman field $\mathbf{B} = B\hat{z}$ was chosen to point to the $+\hat{z}$ direction in the original frame. The emergent dynamics of magnons proves to be more complex than it was for electrons, due to the appearance of $\kappa\mathscr{R}[\hat{e}_\mu \times \mathbf{n}]$ term in the rotated Hamiltonian.

Having completed the rotation to the local basis, the new ground state has $\mathbf{n} = (0, 0, 1)$ everywhere. The HP technique transforms the spin vector into boson field operators with the formula:

$$
n^x = \sqrt{\frac{S}{2}}(e^{i\phi}b^\dagger + e^{-i\phi}b), \quad n^y = i\sqrt{\frac{S}{2}}(e^{i\phi}b^\dagger - e^{-i\phi}b), \quad n^z = S - b^\dagger b, \tag{6.7}
$$

where higher-order terms in the boson field have been neglected. The slightly unorthodox substitution formula used here (which includes the phase factors $e^{\pm i\phi}$) is intended to simplify certain expressions in the final Hamiltonian. The phase ϕ refers, as we recall, to the azimuthal angle of the ground state spin configuration. Substituting (6.7) into (6.6) generates the magnon Hamiltonian

$$
\begin{aligned}
\frac{H[b^\dagger, b]}{SJ} &= \sum_\mu [(\partial_\mu - ia_\mu^3)b^\dagger][(\partial_\mu + ia_\mu^3)b] + m^2 b^\dagger b + \frac{1}{2}\Delta b^2 + \frac{1}{2}\Delta^*(b^\dagger)^2 \\
\text{where} \quad & m^2 - \mathbf{B}\cdot\mathbf{n}_0 - \frac{1}{2}\sum_\mu [(a_\mu^1)^2 + (a_\mu^2)^2], \\
& \Delta = \frac{1}{2}\sum_\mu (a_\mu^1 + ia_\mu^2)^2.
\end{aligned} \tag{6.8}
$$

This is the quadratic magnon Hamiltonian, descended from the HDMZ spin Hamiltonian, in its full generality. The three-dimensional space was assumed throughout the derivation.

The specifics of the ground state spin structure $\mathbf{n}_0$ are encoded through the gauge fields $\mathbf{a}_\mu = (a^1_\mu, a^2_\mu, a^3_\mu)$. After some lengthy algebraic manipulation, one can derive how the emergent gauge fields are related to the physical ground state spin:

$$\begin{aligned} a^1_\mu + i a^2_\mu &= (\hat{\theta} + i\hat{\phi}) \cdot (\partial_\mu \mathbf{n}_0 - \kappa \hat{e}_\mu \times \mathbf{n}_0) \\ a^3_\mu &= -\cos\theta\, \partial_\mu \phi + \kappa (\mathbf{n}_0)_\mu . \end{aligned} \tag{6.9}$$

It is instructive to compare the first line to (5.11). There, the $a^1_\mu + i a^2_\mu$ term measured the twist of the local magnetization $\partial_\mu \mathbf{n}_0$, projected onto the tangent plane $\hat{\theta} + i\hat{\phi}$. We do find a similar expression here, except that the derivative ∂_μ becomes the covariant derivative $\partial_\mu - \kappa \hat{e}_\mu \times$ due to the DM interaction.[2] Starting from the definition of $a^1_\mu + i a^2_\mu$, the mass and anomalous terms in (6.8) can both be written in the physically transparent way

$$\begin{aligned} m^2 &= \mathbf{B} \cdot \mathbf{n}_0 - \frac{1}{2} \sum_\mu (\partial_\mu \mathbf{n}_0 - \kappa \hat{e}_\mu \times \mathbf{n}_0)^2 \\ \Delta &= \frac{1}{2} \sum_\mu [(\hat{\theta} + i\hat{\phi}) \cdot (\partial_\mu \mathbf{n}_0 - \kappa \hat{e}_\mu \times \mathbf{n}_0)]^2 . \end{aligned} \tag{6.10}$$

In fact, the m^2 term is just the negative of the ground state's energy density.

The role of the emergent vector potential is played by the third component $\mathbf{a}^3$ (we are writing $(a^3_x, a^3_y, a^3_z) = \mathbf{a}^3$). The emergent magnetic field arises from taking its curl,[3]

$$(\nabla \times \mathbf{a}^3)_\lambda = \frac{1}{2} \varepsilon_{\lambda\mu\nu} \mathbf{n}_0 \cdot (\partial_\mu \mathbf{n}_0 \times \partial_\nu \mathbf{n}_0) + \kappa (\nabla \times \mathbf{n}_0)_\lambda . \tag{6.11}$$

Inspection of the right-hand side tells us that the emergent magnetic field experienced by the magnons is not only twice what it was for electrons, but also has an additional component proportional to the magnetic vorticity.

The derivation of the generalized magnon Hamiltonian (6.8) and its application to the magnon-skyrmion scattering problem can be found in Refs. [1–4].

6.2 Magnon Excitations of a Spin Spiral

After the hard work of deriving the general magnon Hamiltonian (6.8), one would like to apply it to a few interesting situations. As the first such application, we consider a spiral ground state with the wavevector $\mathbf{k} = \kappa \hat{z}$:

---

[2]Another way to introduce HP bosons in terms of the local triad $(\mathbf{n}_0, \hat{\theta}, \hat{\phi})$ is to write $S^+ = \mathbf{S} \cdot (\hat{\theta} + i\hat{\phi}) = \sqrt{2S} b$, $S^- = \mathbf{S} \cdot (\hat{\theta} - i\hat{\phi}) = \sqrt{2S} b^\dagger$, and $\mathbf{S} \cdot \mathbf{n}_0 = S - b^\dagger b$.

[3]The singular part $\cos\theta (\nabla \times \nabla\phi)_\lambda$ has been dropped from the curl.

$$\mathbf{n}_0 = (\cos\kappa z, \sin\kappa z, 0). \tag{6.12}$$

The gauge fields for this ground state are

$$\begin{aligned} a^1_\mu + ia^2_\mu &= \kappa(\sin\kappa z, \cos\kappa z, 0), \\ a^3_\mu &= \kappa(\cos\kappa z, \sin\kappa z, 0), \end{aligned} \tag{6.13}$$

which yield $m^2 = -\kappa^2/2$, and $\Delta = \kappa^2/2$. As a result, the magnon Hamiltonian (6.8) becomes

$$\begin{aligned} \frac{H}{SJ} &= |(\partial_x + i\kappa\cos\kappa z)b|^2 + |(\partial_y + i\kappa\sin\kappa z)b|^2 + |\partial_z b|^2 \\ &\quad - \frac{1}{2}\kappa^2 b^\dagger b + \frac{1}{2}\kappa^2[b^2 + (b^\dagger)^2]. \end{aligned} \tag{6.14}$$

The next step is to diagonalize the quadratic Hamiltonian by transforming into Fourier space, but the complicated z-dependent gauge field a^3_μ seems to spoil the effort. One can avoid the complication altogether by assuming that the boson field operators depend only on the z-coordinates,

$$\begin{aligned} b(z) &= \sum_k e^{ikz} b_k \\ b^\dagger(z) &= \sum_k e^{-ikz} b^\dagger_k. \end{aligned} \tag{6.15}$$

In this case, we end up with a much simpler Hamiltonian,

$$\begin{aligned} \frac{H}{SJ} &= |\partial_z b|^2 + \frac{1}{2}\kappa^2 b^\dagger b + \frac{1}{4}\kappa^2[b^2 + (b^\dagger)^2] \\ &\Rightarrow \frac{1}{2}\sum_k \begin{pmatrix} b^\dagger_k & b_{-k} \end{pmatrix} \begin{pmatrix} k^2 + \frac{1}{2}\kappa^2 & \frac{1}{2}\kappa^2 \\ \frac{1}{2}\kappa^2 & k^2 + \frac{1}{2}\kappa^2 \end{pmatrix} \begin{pmatrix} b_k \\ b^\dagger_{-k} \end{pmatrix}. \end{aligned} \tag{6.16}$$

The diagonalization of this Hamiltonian is a characteristic Bogoliubov problem. The energy spectrum is found by solving the coupled equations of motion

$$\begin{aligned} [b_k, H] = E_k b_k &= \left(k^2 + \frac{1}{2}\kappa^2\right) b_k + \frac{1}{2}\kappa^2 b^\dagger{}_k, \\ \left[b^\dagger_{-k}, H\right] = E_k b^\dagger_{-k} &= -\frac{1}{2}\kappa^2 b_k - \left(k^2 + \frac{1}{2}\kappa^2\right) b^\dagger_{-k}. \end{aligned} \tag{6.17}$$

The resulting magnon dispersion for the spiral spin state is found to be

$$E_k = \frac{1}{2} SJk\sqrt{k^2 + \kappa^2}. \tag{6.18}$$

For small wavevectors $k \ll \kappa$, the dispersion is linear, $\omega_k \sim SJ\kappa k/2 = (SD/2)k$, which is characteristic of an antiferromagnet, and reflects the fact that the overall magnetization vanishes on a length scale much larger than the spiral wavelength. For $k \gg \kappa$, the quadratic ferromagnetic dispersion $\omega_k \sim SJk^2/2$ is recovered.

In arriving at the simple dispersion formula (6.18), we had to assume that the wavevector of the mode is along the spiral axis $\hat{z}$. In general, the z-dependent modulation in the magnon Hamiltonian acts like a periodic potential that, via Bloch's theorem, leads to the band structure of magnons. Writing the Hamiltonian (6.14) in expanded form ($SJ \equiv 1$) gives

$$H = (\nabla b^\dagger) \cdot (\nabla b) + \frac{1}{2}\kappa^2 b^\dagger b + +\frac{1}{2}\kappa^2 [b^2 + (b^\dagger)^2] \\ +i\kappa \cos \kappa z(\partial_x b^\dagger - \partial_x b) + i\kappa \sin \kappa z(\partial_y b^\dagger - \partial_y b). \quad (6.19)$$

Expansion of a general magnon field can be performed as

$$b(x, y, z) = \sum_{\mathbf{k}}{}' \sum_n b_{n,\mathbf{k}} e^{in\kappa z + i\mathbf{k}\cdot\mathbf{r}}. \quad (6.20)$$

Due to the Umklapp scattering imposed by the z-dependent periodic potential, the summation over $\mathbf{k} = (k_x, k_y, k_z)$ is restricted to $-\kappa/2 < k_z < \kappa/2$, $-\pi < k_x, k_y < \pi$. The integer index n is used to identify each magnon band. Substituting the expansion (6.20) into the magnon Hamiltonian (6.19) leads to an infinite-dimensional matrix form of the Hamiltonian, which can be diagonalized after truncating the matrix at a suitably large band index number $|n| \le n_{max}$. Further details on the magnon band structure worked out in this way can be found in Ref. [5]. Remarkably, careful inelastic neutron scattering experiments performed in Refs. [5, 6] have been able to identify the five lowest magnon bands in a chiral magnet.

6.3 Magnon Excitations of a Single Skyrmion

In this section we discuss how to work out the magnon excitations of a single anti-skyrmion ($Q_s = -1$)

$$\mathbf{n}_0 = (-\sin[f(r)] \sin\varphi, \sin[f(r)] \cos\varphi, \cos[f(r)]), \quad (6.21)$$

where $r = \sqrt{x^2 + y^2}$ and $\varphi = \arctan(y/x)$. We assume that the radial function satisfies the boundary conditions $f(0) = \pi$ and $f(\infty) = 0$. The emergent gauge fields for such an anti-skyrmion yield

$$m^2 = B\cos f - \frac{1}{2}(f')^2 - \frac{(\sin f)^2}{2r^2} - \kappa\left(f' + \frac{\sin 2f}{2r}\right) - \kappa^2(\sin f)^2,$$
$$\Delta = \frac{1}{2}(f')^2 - \frac{(\sin f)^2}{2r^2} + \kappa\left(f' - \frac{\sin 2f}{2r}\right). \tag{6.22}$$

The emergent magnetic field follows from the definition of $\mathbf{a}^3$ [cf. (6.9)]

$$\mathbf{a}^3 = \left(-\frac{\cos f}{r} + \kappa \sin f\right)\hat{\varphi},$$
$$\nabla_2 \times \mathbf{a}^3 = -\frac{(\cos f)'}{r} + \frac{\kappa}{r}(r\sin f)'. \tag{6.23}$$

The first term on the right-hand side of the second equation gives the total flux of -4π, corresponding to two units of flux quanta. The DM part of the emergent flux $\nabla_2 \times \mathbf{a}^3$ creates modulations in the flux density but does not contribute an extra overall flux, provided that $r\sin f$ converges to zero as $r \to \infty$. Another way to see that the total flux is unaffected by the κ-term is to invoke Stokes' theorem to write

$$\int d^2\mathbf{r}\,(\nabla_2 \times \mathbf{n}_0) = \int d\mathbf{l}\cdot\mathbf{n}_0, \tag{6.24}$$

and then use the fact that far from the skyrmion core, $\mathbf{n}_0$ approaches $\hat{z}$ and remains orthogonal to the line integral vector $d\mathbf{l}$. We conclude that magnons see the skyrmion as a source of flux twice as large as that felt by the electrons. A magnon will therefore scatter off the skyrmion will a greater angle of deflection than an electron of comparable incident energy.

For completeness, we show what happens for the case of skyrmion obtained by stereographic projection, where $\cos f = (r^2 - R^2)/(r^2 + R^2)$ and $\sin f = 2rR/(r^2 + R^2)$. Using the relation $f' = -\sin f/r$ to simplify (6.23), we obtain

$$m^2 = B\frac{r^2 - R^2}{r^2 + R^2} + 4\frac{(\kappa R - 1)R^2}{(r^2 + R^2)^2} - 4\frac{(\kappa R)^2 r^2}{(r^2 + R^2)^2}$$
$$\Delta = -4\frac{\kappa R r^2}{(r^2 + R^2)^2}. \tag{6.25}$$

The mass gap in this case approaches B algebraically (or exponentially, if more realistic skyrmion profile had been used). The hybridization gap Δ approaches zero algebraically (or exponentially, for a more realistic profile). The emergent flux $(\nabla_2 \times \mathbf{a}^3)_z$ is not simply -4π because $r\sin f = 2r^2R/(r^2 + R^2)$ approaches $2R$ rather than 0, as $r \to \infty$. The total flux is modified, $-4\pi \to -4\pi(1 - \kappa R)$. This is, however, an artifact of the unusually long tail implied by the stereographic projection. A more realistic and correct approach is to solve for the radial function $f(r)$ numerically, and then use it to obtain $m^2(r)$ and $\Delta(r)$ in (6.23). They can then be fed into the magnon equation of motion, discussed below, to work out the bound magnon spectrum.

Equations (6.22) and (6.23) provide a complete characterization of the magnon Hamiltonian in a single anti-skyrmion background. In order to completely characterize the magnon dynamics, however, one also needs to consider the way the Lagrangian transforms under the rotation $\mathbf{n} \to \mathscr{R}\mathbf{n}$. The geometric action before the rotation is ($\hbar \equiv 1$)

$$S_{\mathrm{B}} = S \int d^2\mathbf{r} dt (\cos\theta - 1)\partial_t\phi. \tag{6.26}$$

Recall that the geometric action has an alternative expression as a Wess-Zumino form, $\int d^2\mathbf{r} dt \int_0^1 du\ \mathbf{n} \cdot (\partial_t \mathbf{n} \times \partial_u \mathbf{n})$ [cf. (1.65)]. The rotation $\mathbf{n} \to \mathscr{R}\mathbf{n}$ obviously leaves the WZ action invariant, provided the matrix $\mathscr{R}$ is independent of both time (which is the case) and the auxiliary variable u (which can be assumed without loss of generality). Thus the Berry phase action is invariant in any local frame of one's choosing, and (6.26) can equally well be regarded as the action in the rotated frame where $\mathbf{n}_0 = (0, 0, 1)$ everywhere.[4] Small deviations from $\mathbf{n}_0$ are captured by the small polar angle θ, hence

$$(1 - \cos\theta)\partial_t\phi \approx \frac{1}{2}\theta^2\partial_t\phi. \tag{6.27}$$

Referring to the HP formula (6.7), we find

$$b \approx \sqrt{\frac{S}{2}}\theta e^{i\phi}, \quad b^* \approx \sqrt{\frac{S}{2}}\theta e^{-i\phi}. \tag{6.28}$$

The Berry phase action in terms of HP bosons reads

$$S_{\mathrm{B}} \approx i \int d^2\mathbf{r} dt\ b^*\partial_t b \tag{6.29}$$

in the rotated frame (or in any frame for that matter). Combining full magnon Lagrangian

$$L = ib^*\partial_t b - H[b, b^*], \tag{6.30}$$

with the Hamiltonian (6.8), yields the magnon equation of motion ($SJ \equiv 1$)

$$i\partial_t b = -\sum_\mu (\partial_\mu + ia^3_\mu)^2 b + m^2 b + \Delta^* b^*. \tag{6.31}$$

---

[4]The geometric action is basically a collection of actions for each spin. With each spin, the action measures the amount of solid angle traced by that spin over time, which must obviously be the same no matter which coordinate frame is used to compute it.

The time derivative does not pick up a gauge term a_0 since the background field is assumed to be static.

Given the circular symmetry of a single skyrmion, the magnon spectrum in the single skyrmion background is appropriately solved using the cylindrical coordinates. We first write

$$\begin{aligned}\sum_{\mu=1,2}(\partial_\mu + i a_\mu^3)^2 &= \left(\nabla + i\left[-\frac{\cos f}{r} + \kappa \sin f\right]\hat{\varphi}\right)^2 \\ &= \frac{\partial^2}{\partial r^2} + \frac{1}{r}\frac{\partial}{\partial r} + \frac{1}{r^2}\left(\frac{\partial}{\partial \phi} - i\cos f + i\kappa r \sin f\right)^2. \end{aligned} \tag{6.32}$$

Each eigenmode is characterized by a definite orbital quantum number l, which appears in the phase $e^{il\varphi}$. The presence of the phase factor $e^{il\varphi}$ in b implies the presence of a $e^{-il\varphi}$ term due to the anomalous term that mixes b with b^*. Consequently, we must try a mixed solution

$$b_l(r,\varphi,t) = A_l(r)e^{il\varphi - i\omega t} + B_l(r)e^{-il\varphi + i\omega t}. \tag{6.33}$$

Inserting the ansatz and matching terms for each phase factor $e^{\pm i(l\varphi - \omega t)}$ gives the coupled differential equations, $\psi_l = \begin{pmatrix} A_l \\ B_l \end{pmatrix}$,

$$\begin{aligned}&\left(m^2(r) - \frac{1}{r}\frac{\partial}{\partial r}\left(r\frac{\partial}{\partial r}\right) + \frac{1}{r^2}[\cos f - \kappa r \sin f - l\sigma^z]^2\right)\psi_l(r) \\ &\quad + \Delta(r)\sigma^x\psi_l(r) = \omega_l \sigma^z \psi_l(r),\end{aligned} \tag{6.34}$$

with $m^2(r)$ and $\Delta(r)$ given in (6.22). A rigorous numerical solution of the confined magnon modes of a single anti-skyrmion is obtained by first solving for $f(r)$ that determines the skyrmion profile. Once this is known numerically, the above eigenvalue problem can be solved numerically for each angular momentum channel l. Identification of bound magnon modes for each deformation channel l has been carried out in Refs. [1, 3] for realistic material parameters. One such example is shown in Fig. 6.1. At a given magnetic field B, the $l = -2$ elliptic deformation mode ($m = -2$ in Fig. 6.1) of the skyrmion was found to have the lowest excitation energy. The energy of the $l = 0$ breathing mode was found to lie slightly below the magnon continuum. The excitation energy of the $l = -2$ mode was found to vanish when the magnetic field was below a certain threshold, indicating that a $l = -2$ "bi-meron" state might condense below this field strength [7].

In the asymptotic region where $\cos f \to 1$, (6.34) reduces to

$$\left(B - \frac{1}{r}\frac{\partial}{\partial r}\left(r\frac{\partial}{\partial r}\right) + \frac{1}{r^2}[1 - l\sigma^z]^2\right)\psi_l(r) = \omega_l\sigma^z\psi_l(r), \tag{6.35}$$

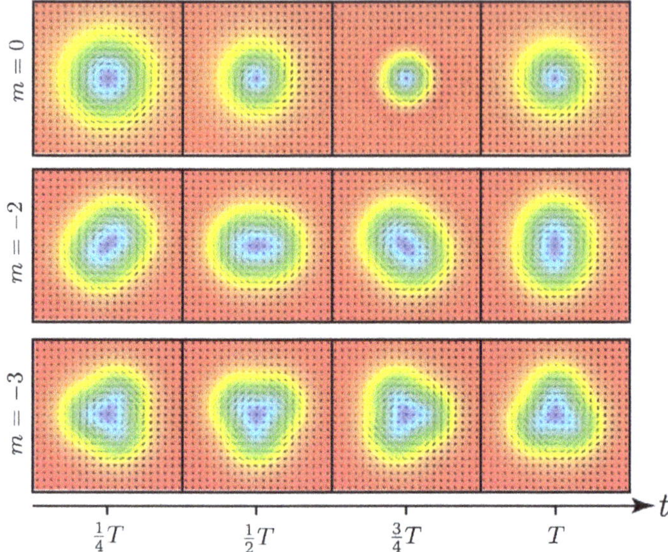

Fig. 6.1 Various deformation modes of a skyrmion [3]. The mode index m is the same as l in (6.34). The gapless $m = -1$ mode which corresponds to the uniform translation of the skyrmion has been omitted. Reprinted with permission from Ref. [3]

which has Bessel function solutions. In particular, the free magnon solution exists only for frequencies $\omega > B$, with the Zeeman energy serving as the magnon gap. This suggests an obvious strategy to tackle the scattering problem, i.e., to separate the full Eq. (6.34) into the sum of a "free part" given in (6.35), and a scattering potential part, which includes the rest of the terms in (6.34). Details of the scattering phase shift and cross-section calculations can be found in Refs. [2, 3]. The magnon-skyrmion scattering problem can also be addressed by carrying out simulations of the Landau–Lifshitz equation, as was performed in Refs. [2, 4]. Seeing the skyrmion as a source of magnetic flux, the magnons scatter preferentially in one direction over the other, resulting in skew scattering and the Hall-like transport. This is also the essence of the Aharonov–Bohm problem of an electron scattering off a localized magnetic flux. Only here, the scattering potential is much more complicated. Multiple rainbow scattering peaks in the differential cross section were reported in Ref. [3].

The other novel aspect of the magnon-skyrmion scattering is the constant drift of the skyrmion due to the incoming magnons, in the direction orthogonal to the magnon wave propagation. This can be understood by going back to the skyrmion equation of motion, which, in the massless limit, yields a skyrmion drift $\dot{\mathbf{R}}$ that is orthogonal to the force. The pressure delivered to the skyrmion by the incoming magnons imparts a force, and the transverse drift velocity. A quantitative analysis of the amount of force delivered by the magnon wave can be found in Refs. [2, 3]. Also, in Sect. 6.6, we will discuss the magnon-induced spin-transfer torque and show that magnons can cause a uniform skyrmion drift in the direction of the magnon flow. A uniform magnon flow can therefore be responsible for both the transverse and longitudinal drift motions of the skyrmion.

6.4 Breathing Modes and Effective Mass

Quite a lot of work went into the theory of bound and scattering magnon modes of a skyrmion in the previous section. In this section, we discuss a simpler and more heuristic way to deduce some of the bound magnon modes for a skyrmion, and use the approach to deduce its effective mass. The mass M_s obtained this way can be used as an input parameter of the skyrmion equation of motion discussed in Chap. 4.

We start by revisiting the model skyrmion configuration

$$\mathbf{n}_s = \left(-\frac{2rR}{r^2+R^2}\sin\varphi, \frac{2rR}{r^2+R^2}\cos\varphi, \frac{r^2-R^2}{r^2+R^2}\right), \tag{6.36}$$

where the azimuthal angle φ comes from $\tan\varphi = y/x$. The contour $r = R$ defines a circular boundary, along which the magnetization lies completely in-plane: $\mathbf{n}(r = R, \varphi) = (-\sin\varphi, \cos\varphi, 0)$. In a fluctuating state, this contour will deform by the amount that generally depends on the angle φ. Instead of writing a constant boundary R, we can construct a function $R(\varphi, t)$ such that the z-component of the spin vanishes at $r = R(\varphi, t)$. Utilizing the 2π azimuthal periodicity, we can write $R(\varphi, t)$ as a Fourier series:

$$R(\varphi, t) = R_s + R_0(t) + R_1(t)e^{i\varphi} + R_{-1}(t)e^{-i\varphi} + \cdots. \tag{6.37}$$

The breathing mode that describes the overall expansion and contraction of the skyrmion is captured by $R_0(t)$. The first harmonics of this mode $R_{\pm 1}(t)$ arises from displacement of the skyrmion center from the origin to $\mathbf{R} = (X, Y)$. A small such displacement results in $\delta R(\varphi, t) = X(t)\cos\varphi + Y(t)\sin\varphi$, or

$$R_1(t) = \frac{1}{2}(X(t) - iY(t)), \quad R_{-1}(t) = \frac{1}{2}(X(t) + iY(t)). \tag{6.38}$$

Have we captured all the small fluctuation modes of a skyrmion? Not quite! The angle of the in-plane magnetization may also deviate from the perfect tangential direction, i.e., the angle φ in (6.36) may be time-dependent. The deformation in the transverse angle can be accounted for by replacing φ with a function,

$$\phi \to \Phi(\varphi, t) = \varphi + \Phi_0(t) + \Phi_1(t)e^{i\varphi} + \Phi_{-1}(t)e^{-i\varphi} + \cdots, \tag{6.39}$$

in (6.36). The "fluctuating skyrmion" ansatz is written as

$$\mathbf{n}_s(t) = \left(-\frac{2rR(t)}{r^2 + R(t)^2} \sin[\Phi(t)], \frac{2rR(t)}{r^2 + R(t)^2} \cos[\Phi(t)], \frac{r^2 - R(t)^2}{r^2 + R(t)^2} \right), \quad (6.40)$$

with $R(t)$ and $\Phi(t)$ as in (6.37) and (6.39).

We substitute $\mathbf{n}_s(t)$ above back into the spin action and work out the Lagrangian in terms of the small deformation modes $R_0(t)$, $\Phi_0(t)$, $R_{\pm 1}(t)$, and $\Phi_{\pm 1}(t)$. As usual, we begin with the geometric phase action:

$$\begin{aligned} S_{\mathrm{B}} &= -\frac{\hbar S}{a^2} \int d^2\mathbf{r} dt (1 - n_z(t)) \partial_t \Phi(t) \\ &= -\frac{2\hbar S}{a^2} \int d^2\mathbf{r} dt \frac{R(t)^2}{r^2 + R(t)^2} \partial_t \Phi(t). \end{aligned} \quad (6.41)$$

The only modes surviving the space integration here are $R_{\mp 1}(t)\partial_t \Phi_{\pm 1}(t)$ and $R_0(t)\partial_t \Phi_0(t)$. A crude calculation provides the approximate action:

$$S_{\mathrm{B}} \approx G R_s \int dt (R_0 \partial_t \Phi_0 + R_1 \partial_t \Phi_{-1} + R_{-1} \partial_t \Phi_1). \quad (6.42)$$

We have used the gyrotropic constant $G = 2hSQ_s/a^2$ with $Q_s = -1$. An equally crude calculation yields the energy functional

$$\begin{aligned} &\int d^2\mathbf{r} dt \left(\frac{J}{2} \sum_{\mu=1,2} (\partial_\mu \mathbf{n}_s(t))^2 + D\mathbf{n}_s(t) \cdot (\nabla \times \mathbf{n}_s(t)) - \mathbf{B} \cdot \mathbf{n}_s(t) \right) \\ &\approx \frac{J}{2} \int dt \left(\Phi_0^2 + \left| \Phi_1 + i \frac{R_1}{R_s} \right|^2 \right). \end{aligned} \quad (6.43)$$

The net effective Lagrangian for the fluctuating skyrmion becomes

$$L_{\mathrm{eff}} \approx G R_s (R_0 \dot{\Phi}_0 + R_1 \dot{\Phi}_{-1} + R_{-1} \dot{\Phi}_1) - \frac{J}{2} \left(\Phi_0^2 + \left| \Phi_1 + i \frac{R_1}{R_s} \right|^2 \right). \quad (6.44)$$

Integrating out the Φ_0 term from the effective Lagrangian yields a contribution $(G^2 R_s^2/2J)\dot{R}_0^2$, while integrating out the $\Phi_{\pm 1}$ term yields $(2G^2 R_s^2/J)\dot{R}_1 \dot{R}_{-1} + iG(\dot{R}_{-1} R_1 - \dot{R}_1 R_{-1})$. All told, we have the effective Lagrangian in terms of R_0 and $R_{\pm 1}$:

$$L_{\mathrm{eff}} = \frac{1}{2} \frac{G^2 R_s^2}{J} (\dot{R}_0^2 + \dot{X}^2 + \dot{Y}^2) + \frac{G}{2} (\dot{X} Y - \dot{Y} X) - \frac{K_0}{2} R_0^2. \quad (6.45)$$

The important lesson to take away from this effective Lagrangian approach is that the mass scales as the area of the skyrmion

$$M_s = \frac{G^2 R_s^2}{J} = \frac{4h^2 S^2 R_s^2}{J a^4} \propto R_s^2. \quad (6.46)$$

The inverse dependence of the mass on the exchange energy J arises from the fact that higher excitation energies are required to create fluctuations if J is larger. In the second-order perturbation theory, the renormalization of physical quantities diminishes inversely with the excitation energy. In the discussion so far, the effect of the chiral DM term (of order D) has not been included explicitly. A numerically small value of $D/J \sim 10^{-1}$ in realistic materials likely implies only a small correction to the above mass estimate. For further details regarding this effective Lagrangian approach and the estimate of the effective mass, readers are directed to Ref. [8].

For skyrmions subject to a rapidly oscillating perturbation, it is more appropriate to generalize the notion of mass parameter as a frequency-dependency quantity. Details on such formulation of the skyrmion mass can be found in Ref. [9].

6.5 Spin Waves in the Skyrmion Lattice

A simplistic treatment of spin excitations in the skyrmion crystal was outlined in Sect. 4.3.2, under the assumption that each skyrmion be treated as a point-like object. The Goldstone mode showed a quadratic dispersion, standing in contrast to the usual linear dispersion of acoustic phonons in solids. This change from a linear to a quadratic dispersion was attributed to the mixing of longitudinal and transverse waves by the Berry phase term.

Here, we discuss a more sophisticated view of the excitations in the skyrmion lattice, in which the whole system is treated as an elastic medium characterized by the displacement vector field $\mathbf{u} = (u_x, u_y)$. The displacement field enters the spin dynamics by the substitution

$$\mathbf{n}(\mathbf{r}, t) = \mathbf{n}_0(\mathbf{r} - \mathbf{u}(\mathbf{r}, t)). \quad (6.47)$$

Figuratively, one can imagine that the ground state spin arrangement $\mathbf{n}_0(\mathbf{r})$ on an elastic piece of rubber, which is deformed slightly in an arbitrary manner. The amount of deformation depends on the location $\mathbf{r}$ and is represented by the vector $\mathbf{u}(\mathbf{r}, t)$, which is time-dependent. The actual value of the spin $\mathbf{n}(\mathbf{r}, t)$ at the deformed location $\mathbf{r}$ is identical to what it was before the deformation took place, $\mathbf{n}_0$, but one has to go back to the original location $\mathbf{r} - \mathbf{u}(\mathbf{r}, t)$ to find it. This is a more sophisticated version of the idea we once employed to derive the skyrmion equation of motion, which was $\mathbf{n}(\mathbf{r}, t) = \mathbf{n}_0(\mathbf{r} - \mathbf{R}(t))$. There, we used a single coordinate $\mathbf{R}(t)$; here, we now have a whole field of displacement vectors. Thus, a more substantial version of the exercise we went through in deriving skyrmion's effective action should give the desired continuum field theory for the displacement field.

Due to the assumption made in (6.47), the Berry phase action is modified to

$$S_{\rm B} = \hbar S \int d^2\mathbf{r}dt \left(\cos[\theta(\mathbf{r} - u(\mathbf{r}, t))] - 1\right)\partial_t\phi(\mathbf{r} - \mathbf{u}(\mathbf{r}, t)). \tag{6.48}$$

We would then consider a Taylor expansion of the angular variables to first order in the displacement, which yields

$$\begin{aligned} \cos[\theta(\mathbf{r} - \mathbf{u}(\mathbf{r}, t))] &\approx \cos[\theta(\mathbf{r})] + \sin\theta(\mathbf{r})[(\mathbf{u}\cdot\nabla)\theta(\mathbf{r})] \\ \phi(\mathbf{r} - \mathbf{u}(\mathbf{r}, t)) &\approx \phi(\mathbf{r}) - (\mathbf{u}\cdot\nabla)\phi(\mathbf{r}). \end{aligned} \tag{6.49}$$

The position-dependent angles $\theta(\mathbf{r})$ and $\phi(\mathbf{r})$ refer now to the ground-state spin orientation. Inserting these Taylor expansions back into the Berry phase action gives

$$S_{\rm B} = -hS\int d^2\mathbf{r}dt\ q_s(u_x\dot{u}_y - u_y\dot{u}_x). \tag{6.50}$$

A straightforward generalization to three-dimensional displacement field $\mathbf{u} = (u_x, u_y, u_z)$ gives the Berry's action $S_{\rm B} = -hS\int d^3\mathbf{r}dt\ \mathbf{q}\cdot(\mathbf{u}\times\dot{\mathbf{u}})$, where $\mathbf{q} = (q_x, q_y, q_z)$ defines the topological density in each plane. Another way to derive the effective action is to make a straightforward adaptation of the formula (5.27):

$$(\delta a_x, \delta a_y, \delta a_0) = 2\pi q_s(u_y, -u_x, u_x\dot{u}_y - u_y\dot{u}_x). \tag{6.51}$$

Equation (6.50) is the elastic-theory version of the skyrmion's geometric action derived in (4.61). The skyrmion density q_s integrates to give the skyrmion charge $\int d^2\mathbf{r}\ q_s = Q_s$.

Having produced the elastic Berry phase action, it is time to rewrite the Hamiltonian in terms of the displacement field. One might imagine doing this by writing $\mathbf{n}(\mathbf{r}, t) = \mathbf{n}_0(\mathbf{r} - \mathbf{u}(\mathbf{r}, t))$ in the HDMZ Hamiltonian and expanding all the terms to first order $\mathbf{n}(\mathbf{r}, t) \approx \mathbf{n}_0 - (\mathbf{u}\cdot\nabla)\mathbf{n}_0$. Although this may be done in principle, it would be a very difficult and confusing task. Instead, one can resort to some symmetry arguments, assuming that the structure about which the deformation is taking place is uniform and rotationally invariant. Strictly speaking, the skyrmion crystal fails to meet either of these requirements, but one could imagine a coarse-grained structure in which skyrmions are uniformly spread out over the whole space. In that setting, the allowed terms that are quadratic in the derivatives of $\mathbf{u}$ are twofold: $(\nabla_2\cdot\mathbf{u})^2$ and $(\nabla_2\times\mathbf{u})^2$. They are the two-dimensional divergence and curl of the deformation field $\mathbf{u}$, respectively. The Lagrangian for the two-dimensional elastic solid, with the Berry phase term included, is

$$L = -hq_sS(\mathbf{u}\times\dot{\mathbf{u}})\cdot\hat{z} + \frac{1}{2}\rho\dot{\mathbf{u}}^2 - \frac{1}{2}\alpha(\nabla_2\cdot\mathbf{u})^2 - \frac{1}{2}\beta(\nabla_2\times\mathbf{u})^2. \tag{6.52}$$

The second order term $\rho\dot{\mathbf{u}}^2/2$ reflects the mass term of the point-particle skyrmion dynamics description, while α and β are two phenomenological constants.

The coupled equation of motion for the components of the deformation field $\mathbf{u}$ follows from this Lagrangian as

$$\begin{aligned} 2hq_s S\dot{u}_y + \rho\ddot{u}_x &= \alpha\partial_x(\nabla\cdot\mathbf{u}) - \beta\partial_y(\nabla\times\mathbf{u}) \\ -2hq_s S\dot{u}_x + \rho\ddot{u}_y &= \alpha\partial_y(\nabla\cdot\mathbf{u}) + \beta\partial_x(\nabla\times\mathbf{u}). \end{aligned} \tag{6.53}$$

Further progress in solving these equations can be made if we assume that the skyrmion density is uniform, $q_s(\mathbf{r}) = q_s$, which is a reasonable assumption for a uniform elastic solid. Adopting the harmonic solution $\mathbf{u}(\mathbf{r},t) = \mathbf{A}e^{i\mathbf{k}\cdot\mathbf{r}-i\omega t}$ with the displacement vector $\mathbf{A} = (A_x, A_y)$ then gives

$$\begin{aligned} -2ihq_s\omega SA_y - \rho\omega^2 A_x &= -\alpha k_x(\mathbf{k}\cdot\mathbf{A}) + \beta k_y(\mathbf{k}\times\mathbf{A}), \\ 2ihq_s\omega SA_x - \rho\omega^2 A_y &= -\alpha k_y(\mathbf{k}\cdot\mathbf{A}) - \beta k_x(\mathbf{k}\times\mathbf{A}). \end{aligned} \tag{6.54}$$

The characteristic equation

$$\left|\begin{pmatrix} \alpha k_x^2 + \beta k_y^2 - \rho\omega^2 & (\alpha-\beta)k_xk_y - 2ihq_sS\omega \\ (\alpha-\beta)k_xk_y + 2ihq_sS\omega & \alpha k_y^2 + \beta k_x^2 - \rho\omega^2 \end{pmatrix}\right| = 0 \tag{6.55}$$

leads to two dispersion branches

$$[\omega_\pm(\mathbf{k})]^2 = \frac{4(2hq_sS)^2 + (\alpha+\beta)\rho\mathbf{k}^2 \pm \sqrt{[4(2hq_sS)^2+(\alpha+\beta)\rho\mathbf{k}^2]^2 - 4\alpha\beta\rho^2\mathbf{k}^4}}{2\rho^2}. \tag{6.56}$$

Let us suppose that the Berry phase term could be dropped, i.e., $q_s = 0$, as for an ordinary two-dimensional solid without topological effects. The two branches $\omega_\pm(\mathbf{k})$ both become gapless in this case, $\omega_l(\mathbf{k}) = \sqrt{\alpha/\rho}|\mathbf{k}|$ (longitudinal) and $\omega_t(\mathbf{k}) = \sqrt{\beta/\rho}|\mathbf{k}|$ (transverse).

Restoring the Berry phase term, we see that one of the excitation branches in (6.56) has a $\mathbf{k} = 0$ gap equal to the cyclotron frequency $\omega_c = 4h|q_s|S/\rho$, while the other branch has a quadratic dispersion,

$$\omega_-(\mathbf{k}) \sim \frac{\sqrt{\alpha\beta}}{4h|q_s|S}\mathbf{k}^2 = \frac{\omega_l(\mathbf{k})\omega_t(\mathbf{k})}{\omega_c}. \tag{6.57}$$

Although the derivation of the gapless dispersion was performed in a very different fashion, it shares the same quadratic dependence on the wavevector we obtained from the point-particle analysis of the phonon mode in Sect. 4.3.2. The essential point in both analyses is that the cyclotron motion induced by the Berry phase effect mixes the longitudinal and transverse phonon modes, resulting in a single gapless mode and one gapped, cyclotron mode. The elastic theory of the magnon modes presented here is adapted from the work of Ref. [10].

Further complexities arising from the skyrmion lattice can be addressed by moving away from the uniform skyrmion density ansatz and imposing harmonic modulations on $q_s(\mathbf{r})$. This procedure will lead to the formation of the Brillouin zone and magnon bands. A full analysis of the magnon band spectrum for the skyrmion lattice was undertaken in a number of papers [11, 12]. Due to the large size of the skyrmion unit cell consisting of many spins, there are many magnon bands forming. Magnon bands in the skyrmion crystal differs significantly from those of the spin spiral worked out in Sect. 6.2 in that the skyrmion magnon bands carry a nonzero Chern number, reflecting the nonzero emergent magnetic field affecting the magnon dynamics. Expected experimental signatures of the topologically nontrivial magnon band structure include the thermal Hall effect [11] and protected chiral edge modes [12].

Finally, we address the question of how these displacement dynamics of the skyrmion lattice affect the electron dynamics through Hund's coupling. The issue was first raised in Sect. 5.3, but was deferred until now to await the proper formulation of the displacement field dynamics. Returning to the strongly Hund-coupled electron Lagrangian,

$$L = \hbar\psi^\dagger[i\partial_t - a_0]\psi - \psi^\dagger\left(\frac{(\mathbf{p}+\hbar\mathbf{a})^2}{2m} + \frac{\hbar^2}{8m}\sum_\mu(\partial_\mu\mathbf{n})^2\right)\psi,$$

we seek to determine the effects of the displacement fluctuation on the electron dynamics that are linear in $\mathbf{u}$. First of all, the correction to a_0 is quadratic in $\mathbf{u}$, as it is proportional to $q\mathbf{u}\times\dot{\mathbf{u}}$ [cf. (6.50)]. The vector potential $\mathbf{a}$ undergoes a linear correction, according to (6.51), as given by

$$\mathbf{a} = \mathbf{a}_0 + 2\pi q_s\mathbf{u}\times\hat{z}. \tag{6.58}$$

The emergent magnetic field is modified by the displacement field to $\mathbf{b} = \mathbf{b}_0(1 - \nabla_2\cdot\mathbf{u})$, which reflects the reduction of the field density due to the uniform expansion of a small patch of area. The Lagrangian itself can be separated into two parts, one that comes from the static gauge fields, and the other part being the "interaction Lagrangian" arising from the displacement field[5]:

$$L_{int.} = -2\pi q_s\mathbf{u}\times\hat{z}\cdot\mathbf{j}^e = -2\pi q_s\hat{z}\cdot(\mathbf{j}^e\times\mathbf{u}). \tag{6.59}$$

We see that the electronic current $\mathbf{j}^e = -i\hbar(\psi^\dagger\nabla\psi - (\nabla\psi^\dagger)\psi)/2m$ couples linearly to the displacement field $\mathbf{u}$, while in the usual phonon problem the coupling is to the gradient of $\mathbf{u}$. This type of electron-phonon coupling was first noted in Ref. [13], and its implications for the electron dynamics have been worked out for two-dimensional skyrmion lattice and three-dimensional columnar skyrmion and monopole lattices in Refs. [14, 15].

[5]The three-dimensional generalization gives $L_{int.} = -2\pi\mathbf{q}\cdot(\mathbf{j}^e\times\mathbf{u})$.

6.6 Magnon-Assisted Skyrmion Motion

Magnons are said to represent a spin-1 excitation as they correspond to the flipping of electron spins from their ground-state orientations. As they propagate, they impart both momentum and spin to the skyrmion with which they interact. The force delivered by the impinging magnons generates a sideways motion of the skyrmion, as discussed in Sect. 6.3. In this section, the focus is on the transfer of angular momentum that takes place during the magnon-skyrmion interaction. We will derive expressions for the magnon-induced spin-transfer torque and modify the Landau–Lifshitz equation to include the magnon effect.

Several different ways to deduce the skyrmion dynamics were discussed in Chap. 4. Among them, the Landau–Lifshitz–Gilbert approach is the one most readily adaptable to the magnon-skyrmion interaction problem. We begin by assuming the continuum HDMZ Hamiltonian and the standard LLG equation

$$\begin{aligned} H_{\text{HDMZ}} &= \frac{J}{2}(\partial_\mu \mathbf{n})^2 + D\mathbf{n}\cdot(\nabla\times\mathbf{n}) - \mathbf{B}\cdot\mathbf{n}, \\ \dot{\mathbf{n}} &= \gamma\mathbf{n}\times\mathbf{f} - \alpha\mathbf{n}\times\dot{\mathbf{n}}. \end{aligned} \tag{6.60}$$

The force part $\mathbf{f} = -\delta H/\delta\mathbf{n} = J\partial_\mu^2\mathbf{n} - 2D\nabla\times\mathbf{n} + \mathbf{B}$ defines the torque

$$\begin{aligned} \mathbf{n}\times\mathbf{f} &= J\mathbf{n}\times\partial_\mu^2\mathbf{n} - 2D\mathbf{n}\times(\nabla\times\mathbf{n}) + \mathbf{n}\times\mathbf{B} \\ &= J\partial_\mu(\mathbf{n}\times\partial_\mu\mathbf{n}) + 2D(\mathbf{n}\cdot\nabla)\mathbf{n} + \mathbf{n}\times\mathbf{B}. \end{aligned} \tag{6.61}$$

We adopt the approach of Ref. [16] and view the spin dynamics as the sum of the slow mode $\mathbf{n}_s$ and a transverse, fast mode parameterized as $\mathbf{n}_f = \mathbf{n}_s \times \mathbf{m}$:

$$\mathbf{n} = (1-\mathbf{m}^2)^{1/2}\mathbf{n}_s + \mathbf{n}_s\times\mathbf{m} \approx \mathbf{n}_s + \mathbf{n}_s\times\mathbf{m}. \tag{6.62}$$

Let us see if some simplifications can be made in the various terms appearing in the torque (6.61) due to such separation of fast and slow modes. One can begin by expressing

$$\mathbf{n}\times\partial_\mu\mathbf{n} \approx \mathbf{n}_s\times\partial_\mu\mathbf{n}_s + \mathbf{n}_f\times\partial_\mu\mathbf{n}_f. \tag{6.63}$$

The cross terms that are linear in $\mathbf{n}_f$ have been neglected, under the assumption that the time average over a full period of the fast motion has been taken. The product of two fast-mode terms appearing at the far right of the above expression can survive the time averaging and should be kept. Simple vector calculus allows us to write

$$\mathbf{n}_f\times\partial_\mu\mathbf{n}_f = \mathbf{n}_s[\mathbf{n}_s\cdot(\mathbf{m}\times\partial_\mu\mathbf{m})] + \mathbf{m}[\mathbf{m}\cdot(\mathbf{n}_s\times\partial_\mu\mathbf{n}_s)]. \tag{6.64}$$

Of the two terms on the right-hand side, the latter is expected to be much smaller than the first because of the assumption $|\partial_\mu\mathbf{n}_s| \ll |\partial_\mu\mathbf{m}|$. One can then drop the second term, to finally reach the approximation

$$\mathbf{n} \times \partial_\mu \mathbf{n} \approx \mathbf{n}_s \times \partial_\mu \mathbf{n}_s + \mathbf{n}_s[\mathbf{n}_s \cdot (\mathbf{m} \times \partial_\mu \mathbf{m})]. \tag{6.65}$$

Due to the extra term on the far right of the above expression, $\mathbf{n} \times \partial_\mu \mathbf{n}$ cannot be simply approximated by replacing $\mathbf{n}$ with the slow mode $\mathbf{n}_s$. No fast-spin contribution survives the time averaging in the Zeeman term, while the DM part can be approximated by

$$(\mathbf{n} \cdot \nabla)\mathbf{n} \approx (\mathbf{n}_s \cdot \nabla)\mathbf{n}_s + \big([\mathbf{n}_s \times \mathbf{m}] \cdot \nabla\big)[\mathbf{n}_s \times \mathbf{m}] \tag{6.66}$$

after taking the time average. The DM-originated torque, proportional to D, can be neglected in comparison to the torque terms in (6.65), which are proportional to J and much larger than D.

If we define the magnonic spin current density as

$$j_\mu^m = J\mathbf{n}_s \cdot (\mathbf{m} \times \partial_\mu \mathbf{m}) \tag{6.67}$$

and assume that it satisfies the steady state condition $\nabla \cdot \mathbf{j}^m = 0$, the series of approximations that led to (6.65) yield the torque

$$\begin{aligned} \mathbf{n} \times \mathbf{f} &= J\partial_\mu(\mathbf{n} \times \partial_\mu \mathbf{n}) + \mathbf{n} \times \mathbf{B} \\ &\approx J\partial_\mu(\mathbf{n}_s \times \partial_\mu \mathbf{n}_s) + \partial_\mu[j_\mu^m \mathbf{n}_s] + \mathbf{n}_s \times \mathbf{B} \\ &\approx \mathbf{n}_s \times \mathbf{f}_s + (\mathbf{j}^m \cdot \nabla)\mathbf{n}_s, \end{aligned} \tag{6.68}$$

where we have introduced the "slow force" $\mathbf{f}_s = -\delta H/\delta \mathbf{n}_s$. The LLG equation for the slow spin variable $\mathbf{n}_s$ (dropping the subscript s) becomes

$$\dot{\mathbf{n}} = \gamma \mathbf{n} \times \mathbf{f} + \gamma\big([\mathbf{j}^m - \mathbf{j}^e] \cdot \nabla\big)\mathbf{n} - \alpha \mathbf{n} \times \dot{\mathbf{n}}. \tag{6.69}$$

The electronic spin-transfer torque is included in this full equation for completeness.

The opposite signs of the electronic ($\mathbf{j}^e$) and the magnonic ($\mathbf{j}^m$) currents have rather counterintuitive consequence. In Sect. 5.2 we showed that skyrmions drift parallel to the electron flow due to the strong Hund's rule coupling. The magnon flow also causes the skyrmion to drift according to the new STT equation (6.69), but in a direction that is opposite to the magnon flow. If we imagine two terminals attached to the ends of a skyrmionic material with the temperature difference across it, both currents $\mathbf{j}^e$ and $\mathbf{j}^m$ will flow from the hot to the cold terminal in obedience of the laws of thermodynamics, but the skyrmion in response to the magnon flow will move from the cold to the hot terminal! As counterintuitive as this may be, numerical simulation of the modified LLG equation indeed confirm this to be the correct picture [16].

As the spin-flip excitation, magnon spins are naturally anti-parallel to the ground state spin orientation. The electron spin, on the other hand, is naturally parallel to the local magnetization due to the strong Hund's coupling effect. This difference helps explain why the two spin current contributions act in opposite manner. One can compare the magnon situation to that of a runner on a treadmill; running forward inevitably results in the reverse motion of the track below.

Direct simulation of the thermally generated magnon flow and its influence on the skyrmion dynamics is possible by adding a stochastic term to the LLG equation. The stochastic LLG equation (or sLLG for short) reads

$$\dot{\mathbf{n}} = -\gamma \mathbf{n} \times (\mathbf{f} + \mathbf{g}) + \alpha \mathbf{n} \times \dot{\mathbf{n}}, \tag{6.70}$$

with the additional stochastic (Langevin) force $\mathbf{g}$ defined by its statistical properties

$$\begin{aligned} \langle \mathbf{g}(\mathbf{r}, t) \rangle &= 0, \\ \langle g_\mu(\mathbf{r}, t) g_\nu(\mathbf{r}', t') \rangle &= \frac{\alpha k_B T}{\gamma} \delta_{\mu\nu} \delta^2(\mathbf{r} - \mathbf{r}') \delta(t - t'). \end{aligned} \tag{6.71}$$

Simulating the temperature difference across the magnetic strip can be achieved with a slow, position-dependent temperature profile $T(x) = T_h + (T_c - T_h)x/L$ across the sample length $0 < x < L$. Because of its stochastic nature, integrating the sLLG equation requires a special trick, as outlined for instance in Ref. [17]. The scheme was extensively applied to study the magnon-driven skyrmion dynamics in Refs. [16, 18]. Magnon-assisted skyrmion transport is expected to hold particular relevance for insulating ferromagnets, since the spin-transfer torque of electronic origin is automatically ruled out in those materials.

Finally, as was discussed in Sect. 6.3, the flow of magnons also generates a transverse force that acts on the skyrmion. Magnons see the skyrmion as the source of an emergent magnetic field and are deflected preferentially to one side of the skyrmion. As a result, the skyrmion experiences a sideways motion that is perpendicular to the magnon flow. This phenomenon was experimentally revealed when the authors of Ref. [19] inadvertently induced a temperature gradient along the radial direction of a small chiral magnetic disk with an electron beam, generating a radial flow of magnons that interacted with the skyrmions inside the magnetic disk. In response, the

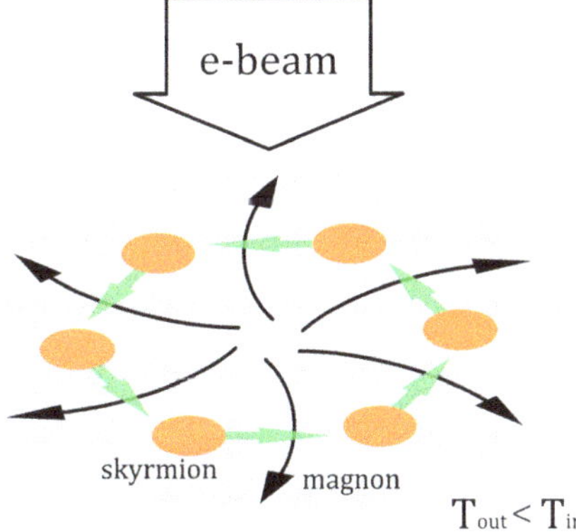

Fig. 6.2 Illustration of skyrmion ratchet motion. The magnons flow from the hot center to the cold boundary. Magnons see the skyrmions as the source of an emergent magnetic field and get deflected. In turn, skyrmions get "kicked back" and set into a rotational motion

skyrmions executed a merry-go-round motion around the disk that was slow enough to be detected by a Lorentz microscope, in a fantastic demonstration of the emergent electrodynamics of magnons. A sketch of the experimental setup is shown in Fig. 6.2.

References

1. Lin, S.Z., Batista, C.D., Saxena, A.: Internal modes of a skyrmion in the ferromagnetic state of chiral magnets. Phys. Rev. B **89**, 024415 (2014)
2. Iwasaki, J., Beckman, A.J., Nagaosa, N.: Theory of magnon-skyrmion scattering in chiral magnets. Phys. Rev. B **89**, 064412 (2014)
3. Schütte, C., Garst, M.: Magnon-skyrmion scattering in chiral magnets. Phys. Rev. B **90**, 094423 (2014)
4. Oh, Y.T., Lee, H., Park, J.H., Han, J.H.: Dynamics of magnon fluid in Dzyaloshinskii-Moriya magnet and its manifestation in magnon-skyrmion scattering. Phys. Rev. B **91**, 104435 (2015)
5. Janoschek, M., Bernlochner, F., Dunsiger, S., Pfleiderer, C., Böni, P., Roessli, B., Link, P., Rosch, A.: Helimagnon bands as universal excitations of chiral magnets. Phys. Rev. B **81**, 214436 (2010)
6. Kugler, M., Brandl, G., Waizner, J., Janoschek, M., Georgii, R., Bauer, A., Seemann, K., Rosch, A., Pfleiderer, C., Böni, P., Garst, M.: Band structure of helimagnons in MnSi resolved by inelastic neutron scattering. Phys. Rev. Lett. **115**, 097203 (2015)
7. Ezawa, M.: Compact merons and skyrmions in thin chiral magnetic films. Phys. Rev. B **83**, 100408(R) (2011)
8. Makhfudz, M., Krüger, B., Tchernyshyov, O.: Inertia and chiral edge modes of a skyrmion magnetic bubble. Phys. Rev. Lett. **109**, 217201 (2012)
9. Schütte, C., Iwasaki, J., Rosch, A., Nagaosa, N.: Inertia, diffusion, and dynamics of a driven skyrmion. Phys. Rev. B **90**, 174434 (2014)
10. Petrova, O., Tchernyshyov, O.: Spin waves in a skyrmion crystal. Phys. Rev. B **84**, 214433 (2011)
11. van Hoogdalem, K.A., Tserkovnyak, Y., Loss, D.: Magnetic texture-induced thermal Hall effects. Phys. Rev. B **87**, 024402 (2013)
12. Roldán-Molina, A., Nunez, A.S., Fernández-Rossier, J.: Topological spin waves in the atomic-scale magnetic skyrmion crystal. New J. Phys. **18**, 045015 (2016)
13. Zang, J., Mostovoy, M., Han, J.H., Nagaosa, N.: Dynamics of skyrmion crystals in metallic thin films. Phys. Rev. Lett. **107**, 136804 (2011)
14. Watanabe, H., Parameshwaran, S.A., Raghu, S., Vishwanath, A.: Anomalous Fermi-liquid phase in metallic skyrmion crystals. Phys. Rev. B **90**, 045145 (2014)
15. Zhang, X.X., Mischenko, A.S., Filippis, G.D., Nagaosa, N.: Electric transport in three-dimensional skyrmion/monopole crystal. Phys. Rev. B **94**, 174428 (2016)
16. Kong, L., Zang, J.: Dynamics of an insulating skyrmion under a temperature gradient. Phys. Rev. Lett. **111**, 067203 (2013)
17. García-Palacios, J.L., Lázaro, F.J.: Langevin-dynamics study of the dynamical properties of small magnetic particles. Phys. Rev. B **58**, 14937 (1998)
18. Lin, S.Z., Batista, C.D., Reichhardt, C., Saxena, A.: AC current generation in chiral magnetic insulators and skyrmion motion induced by the spin Seebeck effect. Phys. Rev. Lett. **112**, 187203 (2014)
19. Mochizuki, M., Yu, X.Z., Seki, S., Kanazawa, N., Koshibae, W., Zang, J., Mostovoy, M., Tokura, Y., Nagaosa, N.: Thermally driven ratchet motion of a skyrmion microcrystal and topological magnon Hall effect. Nat. Mat. **13**, 241 (2014)

Chapter 7
Miscellaneous Topics

This chapter deals with a number of themes that are useful to know, but which are not necessarily correlated with one another. Some of these topics are concerned with exploring the technological potential of "engineering" skyrmions through nucleation and tunneling processes. Several physical mechanisms capable of nucleating individual skyrmions are considered, ranging from electrical currents to optical beams. Furthermore, a minimal model of a multiferroic skyrmion insulator is introduced. We begin the chapter with a formal discussion of the symmetries of the minimal Lagrangian for chiral magnets and the various conservation laws which follow from Noether's theorem.

7.1 Symmetry Consideration

This section is devoted to the close examination of the continuous symmetries inherent in the minimal HDMZ model of chiral magnets. The HDMZ Lagrangian is

$$\begin{aligned} L &= L_{\rm B} - H \\ L_{\rm B} &= -S(1-\cos\theta)\partial_t\phi = -S\mathbf{A}\cdot\dot{\mathbf{n}} \\ H &= \frac{J}{2}(\partial_\mu \mathbf{n})^2 + D\mathbf{n}\cdot(\nabla\times\mathbf{n}) - \mathbf{B}\cdot\mathbf{n}, \end{aligned} \tag{7.1}$$

with the implicit summation over the spatial indices μ.[1] Our discussion of the symmetries of this model is broken into two parts: the spacetime translation symmetry and the rotational symmetry about the magnetic field. We follow the symmetry arguments for the HDMZ Hamiltonian presented in Ref. [1].

[1]In this section, we use α, β to refer to spacetime indices, and μ, ν, λ to refer to space indices.

J.H. Han, *Skyrmions in Condensed Matter*, Springer Tracts
in Modern Physics 278, https://doi.org/10.1007/978-3-319-69246-3_7

7.1.1 Spacetime Translation Symmetry

The HDMZ action remains invariant with respect to the uniform translation of the coordinates $\mathbf{r} \to \mathbf{r} + \delta\mathbf{r}$, provided that the Zeeman field is uniform $\mathbf{B}(\mathbf{r}) = \mathbf{B}$. For the Hamiltonian that is not explicitly time-dependent, the action is also invariant under the time translation $t \to t + \delta t$. The variation of the HDMZ action with respect to $\delta\mathbf{r}$ and δt can be calculated using $\delta\mathbf{n} = (\delta\mathbf{r} \cdot \nabla + \delta t \partial_t)\mathbf{n} = \delta x_\alpha \partial_\alpha \mathbf{n}$, where α runs over both time and space coordinates:

$$\delta S = \delta x_\beta \int d^2\mathbf{r} dt \; \partial_\alpha \left(\delta_{\alpha\beta} L - \frac{\partial L}{\partial(\partial_\alpha \mathbf{n})} \cdot \partial_\beta \mathbf{n} \right) = 0. \tag{7.2}$$

The conservation law following from this spacetime translation symmetry is $\partial_\alpha T_{\alpha\beta} = 0$, where the definition of the energy-momentum tensor is given by

$$T_{\alpha\beta} = \frac{\partial L}{\partial(\partial_\alpha \mathbf{n})} \cdot \partial_\beta \mathbf{n} - \delta_{\alpha\beta} L. \tag{7.3}$$

We have seen a related expression in Sect. 4.2.3, where the indices of the energy-momentum tensor covered only the spatial components [cf. (4.86)]. The energy-momentum tensor there arose from attempts to write down the continuity equation for the topological density, not from the direct consideration of the spacetime symmetry. Moreover, the temporal component was not a part of the consideration.

It is worth examining the consequences of the conservation law $\partial_\alpha T_{\alpha\beta} = 0$ separately when the second index β is temporal, and when it is spatial. When $\beta = t$, the continuity equation reads $\partial_0 T_{00} + \partial_\mu T_{\mu 0} = 0$.[2] From the definition of $T_{\alpha\beta}$ it is easy to see that

$$T_{\mu 0} = -\frac{\partial H}{\partial(\partial_\mu \mathbf{n})} \cdot \dot{\mathbf{n}}. \tag{7.4}$$

Next, to determine T_{00} one needs to know what $\partial L / \partial \dot{\mathbf{n}}$ is. Since the Hamiltonian does not contain $\dot{\mathbf{n}}$, and the geometric phase is the only piece of the action that contains $\dot{\mathbf{n}}$ as $L_{\mathrm{B}} = -S\mathbf{A} \cdot \dot{\mathbf{n}}$ [cf. (1.65)], we have

$$\frac{\partial L}{\partial \dot{\mathbf{n}}} = -S\mathbf{A}, \tag{7.5}$$

where $\mathbf{A}$ stands for the vector potential of a "magnetic monopole" of unit charge $\nabla_{\mathbf{n}} \times \mathbf{A} = \mathbf{n}$. A detailed discussion of $\mathbf{A}$ and its curl was provided in Sect. 1.4.3. It follows that $T_{0\beta} = -S\mathbf{A} \cdot \partial_\beta \mathbf{n} - \delta_{0\beta} L$, or

$$T_{00} = H, \quad T_{0\nu} = -S\mathbf{A} \cdot \partial_\nu \mathbf{n}. \tag{7.6}$$

---

[2] We write the time derivative as ∂_0, ∂_t, or with a dot over the object such as $\dot{\mathbf{n}}$, as the aesthetics of the equation at hand dictates. The temporal component of the tensor will be denoted by 0 or t.

The $\beta = 0$ equation $\partial_\alpha T_{\alpha 0} = 0$ turns out to be a continuity equation for the energy density

$$\partial_t H - \partial_\mu \left(\frac{\partial H}{\partial(\partial_\mu \mathbf{n})} \cdot \dot{\mathbf{n}} \right) = 0. \tag{7.7}$$

Next, we discuss the conservation law where $\beta = \nu$ is one of the spatial indices. First we need to determine the tensor elements where both components are spatial:

$$\begin{aligned} T_{\mu\nu} &= -\frac{\partial H}{\partial(\partial_\mu \mathbf{n})} \cdot \partial_\nu \mathbf{n} + \delta_{\mu\nu} H - \delta_{\mu\nu} L_{\mathrm{B}} \\ &= T^s_{\mu\nu} - \delta_{\mu\nu} L_{\mathrm{B}}. \end{aligned} \tag{7.8}$$

The static energy-momentum tensor $T^s_{\mu\nu}$ is defined in terms of the Hamiltonian H. The same static energy-momentum tensor appeared first in (4.86). Writing out the conservation law $\partial_\alpha T_{\alpha\nu} = \partial_0 T_{0\nu} + \partial_\mu T_{\mu\nu} = 0$ explicitly, with $T_{0\nu}$ from (7.6) and $T_{\mu\nu}$ from (7.8) above, we obtain

$$\partial_\mu T^s_{\mu\nu} = S\big[\partial_\nu(\mathbf{A} \cdot \partial_t \mathbf{n}) - \partial_t(\mathbf{A} \cdot \partial_\nu \mathbf{n})\big]. \tag{7.9}$$

The right-hand side of this equation looks like the curl of some vector. Its meaning becomes clear once we write the vector potential explicitly,

$$\mathbf{A} = -\frac{1 + \cos\theta}{\sin\theta} \hat{\phi}. \tag{7.10}$$

Since $\partial_\nu \mathbf{n} = \hat{\theta} \partial_\nu \theta + \hat{\phi} \sin\theta \partial_\nu \phi$, the inner product $\mathbf{A} \cdot \partial_\nu \mathbf{n}$ yields

$$\mathbf{A} \cdot \partial_\nu \mathbf{n} = -(1 + \cos\theta) \partial_\nu \phi = 2 a_\nu, \tag{7.11}$$

which is twice the emergent gauge field a_ν. In other words, we have the neat relation

$$T_{0\nu} = -S\mathbf{A} \cdot \partial_\nu \mathbf{n} = -2S a_\nu \tag{7.12}$$

and (7.9) becomes

$$\partial_\mu T^s_{\mu\nu} = 2S(\partial_\nu a_t - \partial_t a_\nu) = S\mathbf{n} \cdot (\partial_\nu \mathbf{n} \times \partial_t \mathbf{n}). \tag{7.13}$$

Going back to the discussion in Sect. 4.2.3 where the topological current was defined as

$$J_{\mu\nu} = \varepsilon_{\mu\nu\lambda} (\partial_\lambda \mathbf{n} \times \partial_t \mathbf{n}) \cdot \mathbf{n} = \varepsilon_{\mu\nu\lambda} \partial_\rho T^s_{\rho\lambda}, \tag{7.14}$$

we realize that (7.13) is the same as (7.14) once the antisymmetric tensor $\varepsilon_{\mu\nu\lambda}$ is removed from both sides.

In brief, time translation invariance of the HDMZ action gave the energy conservation law, as embodied in the continuity equation for the energy density (7.7). Since we know that the invariance under the spatial translation should be related to the momentum conservation, we may interpret the right-hand side of (7.13) as the time derivative of the momentum density of spins in the ν-direction, and the left-hand side of (7.13) as the spatial divergence of the momentum current.

7.1.2 Rotational Symmetry

Next, we consider the symmetry of the HDMZ action under the rotation of the spin about the magnetic field direction $\mathbf{B} \parallel \hat{z}$. An infinitesimal rotation by (time-dependent) ω about the $\hat{z}$-axis can be expressed as

$$\delta\mathbf{n} = \omega\hat{z} \times \mathbf{n} \quad (|\omega| \ll 1), \tag{7.15}$$

and feeding $\delta\mathbf{n}$ back into the variation of the action δS_{HDMZ} gives

$$\begin{aligned} \delta S_{\text{HDMZ}} &= \int \delta\mathbf{n} \cdot \left[\frac{\partial L}{\partial \mathbf{n}} - \partial_\alpha \left(\frac{\partial L}{\partial(\partial_\alpha \mathbf{n})} \right) \right] \\ &= \int \omega\hat{z} \cdot \mathbf{n} \times \left[\frac{\partial L}{\partial \mathbf{n}} - \partial_\alpha \left(\frac{\partial L}{\partial(\partial_\alpha \mathbf{n})} \right) \right]. \end{aligned} \tag{7.16}$$

Setting the variation equal to zero results in an equation of motion for the spin. Since $L = L_{\text{B}} - H_{\text{HDMZ}}$, the contribution of each term to the equation of motion can be examined separately. Let us begin with the Hamiltonian part:

$$-\frac{\partial H_{\text{HDMZ}}}{\partial \mathbf{n}} + \partial_\mu \left(\frac{\partial H_{\text{HDMZ}}}{\partial(\partial_\mu \mathbf{n})} \right) = J\partial_\mu^2 \mathbf{n} - 2D\nabla \times \mathbf{n} + \mathbf{B}. \tag{7.17}$$

Operating from the left with $\omega\hat{z} \cdot \mathbf{n} \times$ returns the spatial part of the variation δS_{HDMZ}:

$$\omega\hat{z} \cdot \left[J\mathbf{n} \times \partial_\mu^2 \mathbf{n} - 2D\mathbf{n} \times (\nabla \times \mathbf{n}) \right] = \omega\left(\nabla \cdot \mathbf{j}^{n_z} + 2D(\mathbf{n} \cdot \nabla)n^z \right), \tag{7.18}$$

where we introduced the spin current projected onto the $\hat{z}$-direction $j_\mu^{n_z} = J\hat{z} \cdot [\mathbf{n} \times \partial_\mu \mathbf{n}]$.

The temporal part of the variation comes from L_{B} only, and is easiest to work out when it is written in the way $L_{\text{B}} = S(\cos\theta - 1)\partial_t\phi$:

$$\int \delta L_{\text{B}} = S \int (\cos\theta - 1)\partial_t \omega = -S \int \omega\partial_t[\cos\theta] = -S \int \omega\partial_t n^z. \tag{7.19}$$

Putting the temporal and spatial pieces together, one arrives at the Landau–Lifshitz equation projected onto the z-axis:

$$S\partial_t n^z = \nabla \cdot \mathbf{j}^{n_z} + 2D(\mathbf{n}\cdot\nabla)n^z. \tag{7.20}$$

This is "almost" a continuity equation, were it not for the DM term at the far right. The origin of the nonconservation is easy to understand; the DM interaction $\mathbf{n}\cdot\nabla\times\mathbf{n}$ is invariant only under the *simultaneous rotation* of spin $\mathbf{n}$ and space $\mathbf{r}$ by the same orthogonal matrix $\mathcal{R}$. The variation we are considering here only rotates the spin.

This time, we consider the variation of the action under the *spatial rotation* of the coordinates $\mathbf{r}$. We will be met with another failed attempt to derive a conservation law, only this time it will be for the orbital angular momentum. Sewing the two failures together will finally mend the problem, and yield the law of the total angular momentum conservation.

A spatial rotation about the z-axis by an infinitesimal angle ω yields the displacement $\delta\mathbf{r} = \omega\hat{z}\times\mathbf{r}$, and the change in the spin vector given by

$$\delta\mathbf{n} = \mathbf{n}(\mathbf{r}+\delta\mathbf{r}) - \mathbf{n}(\mathbf{r}) = (\delta\mathbf{r}\cdot\nabla)\mathbf{n} = \omega\hat{z}\cdot(\mathbf{r}\times\nabla)\mathbf{n}. \tag{7.21}$$

Inserting this into the first line of (7.16) gives

$$\delta S_{\text{HDMZ}} = \int \omega(x\partial_y\mathbf{n} - y\partial_x\mathbf{n})\cdot\left(\frac{\partial L}{\partial\mathbf{n}} - \partial_\alpha\left(\frac{\partial L}{\partial(\partial_\alpha\mathbf{n})}\right)\right), \tag{7.22}$$

and some simple algebra lets us write

$$\begin{aligned}\partial_x\mathbf{n}\cdot\frac{\partial L}{\partial\mathbf{n}} - \partial_x\mathbf{n}\cdot\partial_\alpha\left(\frac{\partial L}{\partial(\partial_\alpha\mathbf{n})}\right) &= \partial_x L - \partial_\alpha\left(\partial_x\mathbf{n}\cdot\frac{\partial L}{\partial(\partial_\alpha\mathbf{n})}\right)\\ &= -\partial_\alpha T_{\alpha x}.\end{aligned} \tag{7.23}$$

With this expression, one can rewrite δS_{HDMZ} in (7.22) in the compact form

$$\begin{aligned}\delta S_{\text{HDMZ}} &= -\int \omega\left(x\partial_\alpha T_{\alpha y} - y\partial_\alpha T_{\alpha x}\right)\\ &= -\int \omega\Big(\partial_\alpha[xT_{\alpha y} - yT_{\alpha x}] + T_{yx} - T_{xy}\Big).\end{aligned} \tag{7.24}$$

As the expression inside the parenthesis vanishes for physical trajectories, we are left with

$$\partial_\alpha[xT_{\alpha y} - yT_{\alpha x}] = T_{xy} - T_{yx}, \tag{7.25}$$

$\alpha = t, x, y, z$. Again, this almost has the appearance of a conservation law, which is broken by the nonzero value of $T_{xy} - T_{yx} = -T^s_{xy} + T^s_{yx}$, where $T_{xy} = -T^s_{xy}$ follows from (7.8).

Further inspection of the Hamiltonian reveals that the asymmetry $T^s_{xy} \neq T^s_{yx}$ arises from the DM interaction. Working through some algebra, the right-hand side of (7.25) becomes

$$\begin{aligned} -T^s_{xy} + T^s_{yx} &= \frac{\partial H}{\partial(\partial_y \mathbf{n})} \cdot \partial_x \mathbf{n} - \frac{\partial H}{\partial(\partial_x \mathbf{n})} \cdot \partial_y \mathbf{n} \\ &= 2D\Big(\big(n^x \partial_x + n^y \partial_y\big)n^z - D\big(\partial_x[n^z n^x] + \partial_y[n^z n^y]\big)\Big). \end{aligned} \quad (7.26)$$

In order to gain some insight into the left-hand side of (7.25), we first focus on its temporal component

$$l^z = xT_{0y} - yT_{0x} = -2S(xa_y - ya_x). \quad (7.27)$$

Here, the association $T_{0\nu} = -2Sa_\nu$ comes from (7.12). The label l^z reflects our attempt to establish $-2S\mathbf{r} \times \mathbf{a}$ as the orbital angular momentum density.

To digress briefly, we note that integrating l^z over the two-dimensional space gives

$$\begin{aligned} L^z = \int d^2\mathbf{r}\, l^z &= -S \int d^2\mathbf{r}[\partial_x(x^2)a_y - \partial_y(y^2)a_x] \\ &= S \int d^2\mathbf{r}\, [x^2 \partial_x a_y - y^2 \partial_y a_x] \\ &= S \int d^2\mathbf{r}\, (x^2 + y^2)(\partial_x a_y - \partial_y a_x) \\ &= 2\pi S \int d^2\mathbf{r}\, \mathbf{r}^2 q_s(\mathbf{r}). \end{aligned} \quad (7.28)$$

We thus see that the orbital angular momentum is related to the weighted moment of the skyrmion density $q_s = (\boldsymbol{\nabla}_2 \times \mathbf{a})/2\pi$.

We are now ready to recast (7.25) as the equation of motion for l^z

$$\partial_t l^z + \boldsymbol{\nabla} \cdot \mathbf{j}^{l_z} = 2D(\mathbf{n} \cdot \boldsymbol{\nabla}_2)n^z, \quad (7.29)$$

with the orbital angular momentum current defined as

$$\mathbf{j}^{l_z} = (xT_{xy} - yT_{xx} + Dn^x n^z,\ xT_{yy} - yT_{yx} + Dn^y n^z). \quad (7.30)$$

As expected earlier, the orbital angular momentum is not conserved. Now, by combining the orbital equation (7.29) together with the spin equation (7.20), one can arrive at a proper continuity equation for the total angular momentum in two dimensions:

$$\partial_t(l^z - Sn^z) + \boldsymbol{\nabla} \cdot (\mathbf{j}^{l_z} + \mathbf{j}^{n_z}) = 0. \quad (7.31)$$

The hefty exercise we went through offers more than some aesthetic satisfaction. The total angular momentum (density) is now properly identified with $l^z - Sn^z$, where the appearance of S accounts for the fact that the size of spin S matters in quantifying the spin angular momentum, and the minus sign arises from the negative value of the gyromagnetic ratio.

Having endured these formal calculations, let us now try to determine how much orbital angular momentum is carried by a single anti-skyrmion with the functional form

$$\mathbf{n}_s = (-\sin[f(r)]\sin\varphi, \sin[f(r)]\cos\varphi, \cos[f(r)]) . \tag{7.32}$$

The topological density $q_s = \mathbf{n}_s \cdot (\partial_x \mathbf{n}_s \times \partial_y \mathbf{n}_s)/4\pi$ for this spin configuration works out to $q_s = -(\cos f)'/4\pi r = -(n^z)'/4\pi r$, and the associated orbital angular momentum is

$$L^z = 4\pi^2 S \int_0^\infty r dr\, r^2 q_s(r) = -\pi S \int_0^\infty dr\, r^2 (n^z - 1)', \tag{7.33}$$

where we used the identity $(\cos f)' = (n^z)' = (n^z - 1)'$ in the final expression. Comparing this to the total spin angular momentum of the skyrmion,

$$\begin{aligned} M^z &= -S \int d^2\mathbf{r}\, (n^z - 1) \\ &= -2\pi S \int_0^\infty dr\, r(n^z - 1) = \pi S \int_0^\infty r^2 (n^z - 1)', \end{aligned} \tag{7.34}$$

we arrive at the interesting conclusion that the total angular momentum of the skyrmion is zero:

$$J^z = L^z + M^z = 0! \tag{7.35}$$

One can interpret this result as follows. When there is no skyrmion present and all the spins are ferromagnetically aligned, the total angular momentum equals some constant which we can take to be zero; as the skyrmion forms, the magnetic structure develops a certain amount of orbital angular momentum and an exactly opposite amount of spin angular momentum, in a manner that conserves the total angular momentum.

7.2 Theory of Skyrmion Generation

The enthusiasm for skyrmion physics is in large part derived from their potential to act as mobile carriers of information bits. An essential component of the program to turn this idea into a working scheme is the ability to generate and destroy individual skyrmions at will. The LLG equation analysis of the topological charge dynamics in Sect. 4.2.3 demonstrated that the total skyrmion number is a constant of motion, due to the continuity equation for the skyrmion density [cf. (4.79)]. Within the continuum theory of spin dynamics, skyrmions can only be created in pairs, or not at all.

It should be emphasized, though, that the argument holds only for the continuum model of a chiral magnet, and in the absence of coupling to external sources such as electromagnetic radiation or electric current. In this section, we will discuss how the introduction of these external agents can help in nucleating individual skyrmions. We note in advance that neither method relies on the application of external magnetic field, which tends to create skyrmions *en masse* rather than one object at a time.

7.2.1 Electrical Means of Skyrmion Generation

As we have seen earlier, the spin-transfer torque coupling of magnetic moments to the electron spin modifies the LLG equation to [cf. (6.69)]

$$\dot{\mathbf{n}} = \gamma \mathbf{n} \times \mathbf{f} - \left(\mathbf{j}^e \cdot \nabla\right)\mathbf{n} - \alpha \mathbf{n} \times \dot{\mathbf{n}}.$$

Working out the time derivative of $b_\mu = (1/2)\varepsilon_{\mu\nu\lambda}\mathbf{n} \cdot (\partial_\nu \mathbf{n} \times \partial_\lambda \mathbf{n})$ in the presence of the modified LLG equation above, one finds that the previous topological conservation law (4.79) is modified to[3]

$$\begin{aligned} \dot{b}_\mu + \partial_\nu J_{\mu\nu} &= (\mathbf{b} \cdot \nabla) j^e_\mu \\ J_{\mu\nu} &= \varepsilon_{\mu\nu\lambda}\partial_\lambda \mathbf{n} \cdot [\gamma \mathbf{f} - \alpha \mathbf{n} \times \mathbf{f}] + b_\mu j^e_\nu. \end{aligned} \tag{7.36}$$

We see that the modified definition of the topological current tensor includes the additional term $q_\mu j^e_\nu$. More importantly, the topological density is no longer conserved, due to the $(\mathbf{q} \cdot \nabla) j^e_\mu$ term on the right-hand side of (7.36). Here, $\mathbf{b} = (b_x, b_y, b_z)$ describes the emergent magnetic field in each plane, and the equation implies that an inhomogeneous current flow in the direction perpendicular to the plane effectively can change the skyrmion number within that plane. To be more specific, integrating (7.36) over the whole three-dimensional space gives

$$\dot{Q}_\mu = \frac{1}{2\pi} \int d^3\mathbf{r}\, (\mathbf{b} \cdot \nabla) j^e_\mu = - \int d^3\mathbf{r}\, j^e_\mu (\nabla \cdot \mathbf{b}). \tag{7.37}$$

A current flow in the μ-direction is a necessary requirement to induce the generation of skyrmions in a plane normal to the current.

---

[3]The modification of the continuity equation for the topological density to include electronic coupling does not appear to have been discussed in the literature. The author derived the following modified continuity equation during the writing of this book and bears sole responsibility for its validity.

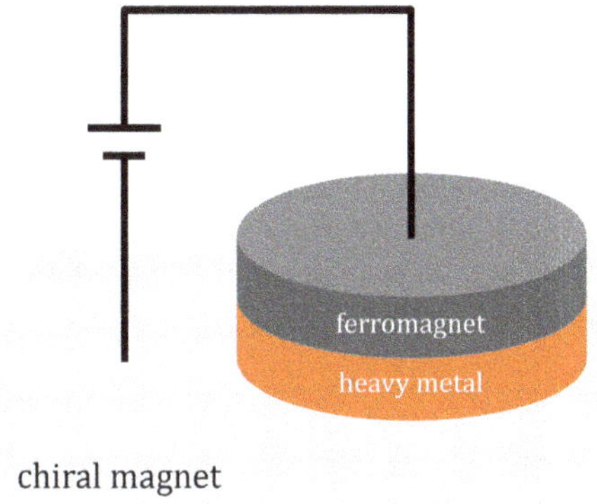

Fig. 7.1 Possible geometry for the efficient production of skyrmions. A ferromagnetic/heavy metal layer is deposited on top of a chiral magnet in order to induce a vertical flow of spin-polarized current. According to (7.37), such geometry is conducive to nucleating skyrmions in the plane of the chiral magnet

One can therefore imagine a pillar geometry, as in Fig. 7.1, where a chiral magnetic metal layer and a ferromagnetic film (e.g., Co) are separated by a nonmagnetic metallic spacer layer, e.g., Cu (or an even heavier metal if a larger interfacial DM interaction is preferred). The bias applied between the top and bottom layers will induce a spin-polarized current flow that could lead to the nucleation of skyrmions by the spin-transfer torque effect, without the need to apply an external magnetic field. A number of numerical simulations have reported the successful nucleation of skyrmions through the vertical injection of current [2–5], in support of this prediction.

The interaction with the electronic current adds a term

$$H_{int} = \int d^3\mathbf{r}\, \mathbf{a} \cdot \mathbf{j}^e \tag{7.38}$$

to the magnetic Hamiltonian. The awkward presence of the emergent vector potential $\mathbf{a}$ (which is hard to write in terms of $\mathbf{n}$) can be remedied when the current $\mathbf{j}^e$ is assumed stationary. In this case, $\mathbf{j}^e$ can be written in terms of another vector field $\mathbf{j}^e = \nabla \times \mathbf{c}$ (we are treating $\mathbf{j}^e$ as a classical vector), and after integrating by parts, the electronic current coupling becomes

$$H_{int} = \int d^3\mathbf{r}\, \mathbf{b} \cdot \mathbf{c}. \tag{7.39}$$

As the emergent magnetic field is related to the spin texture by

$$b_\mu = \frac{1}{2} q_\mu = \frac{1}{4} \varepsilon_{\mu\nu\lambda} \mathbf{n} \cdot (\partial_\nu \mathbf{n} \times \partial_\lambda \mathbf{n}), \tag{7.40}$$

we are now able to express H_{int} in terms of the physical spin vector $\mathbf{n}$.

Since a moving electron carries with it a kinetic energy $m_e \mathbf{v}_e^2/2$, the kinetic energy density associated with the finite current $\mathbf{j}^e = n_e \mathbf{v}_e$ is $(K/2)\mathbf{j}_e^2$, where K is some phenomenological constant. One could write down a modified Hamiltonian as

$$H_{\mathbf{n},\,\mathbf{j}^e} = H_{\mathbf{n}} + \mathbf{b}\cdot\mathbf{c} + \frac{K}{2}(\nabla\times\mathbf{c})^2, \tag{7.41}$$

capturing the magnetization dynamics of spins coupled to a stationary flow of current. With such Hamiltonian, one could study the ideal magnetization pattern for a given stationary current $\mathbf{j}^e = \nabla\times\mathbf{c}$.

In a further attempt to develop some insight into the nature of the coupling (7.40), write $\mathbf{c}$ approximately as

$$\mathbf{c} \approx -\frac{1}{2}\mathbf{r}\times\mathbf{j}^e, \tag{7.42}$$

for a slowly varying current $\mathbf{j}^e$. The expression on right-hand side suggests that we can interpret $\mathbf{c}$ as proportional to the orbital angular momentum carried by the current, $\mathbf{r}\times\mathbf{j}^e \sim \mathbf{l}^e$, and the interaction Hamiltonian

$$H_{int} \sim -\int d^3\mathbf{r}\,\mathbf{b}\cdot\mathbf{l}^e \tag{7.43}$$

as encoding the coupling of the emergent magnetic field $\mathbf{b}$ to the orbital angular momentum $\mathbf{l}^e$ carried by the electronic current. Such interpretation suggests that a means of generating electronic vorticity ($\mathbf{l}^e \neq 0$) can also act as a source term for the topological density vector $\mathbf{b}$.

Strictly speaking, it is impossible to have a vertical current flow in a truly two-dimensional geometry and the skyrmion number must be conserved even when the spin-transfer torque coupling is present, at least in the continuum theory where such conservation laws can be physically grounded. In reality, the lattice theory is free from such a strict conservation principle. For instance, the possibility of creating skyrmions by a circularly rotating current in a two-dimensional lattice model was investigated in Ref. [6], and Ref. [7] showed that an electric current moving past a protruding boundary such as the constriction point in a nanowire could act as a source of skyrmion generation.

Most of what we know in regard to the skyrmion generation by an electric current comes from the numerical investigation of the LLG equation. For further details, we direct the interested reader to Refs. [2, 7], where the generation of skyrmions and their transport along a narrow strip was investigated with a view toward racetrack memory applications. Several experimental papers have also reported the nucleation and destruction of skyrmions by a variety of current injection techniques [8–10]. The rough picture emerging from a variety of simulations and experiments is that a strongly inhomogeneous current distribution is conducive to the formation of skyrmions.

7.2.2 Optical Means of Skyrmion Generation

As we have just seen, skyrmion generation by electrical means relies on the spin-transfer torque mechanism. Another interesting proposal is to generate skyrmions using high-intensity circularly polarized light beams focused on a region comparable to the size of a few skyrmions [11]. In this method, the idea is to couple the spins to the time-dependent Zeeman field

$$H_Z(t) = -B_0 n^z - B_1(r)[\cos(\omega t + m\varphi)n^y - \sin(\omega t + m\varphi)n^x], \qquad (7.44)$$

where $\varphi = \arctan(y/x)$ is the azimuthal angle in the plane of the chiral magnet, and $|\omega|$ is the frequency of circularly polarized light. Depending on the sign of ω, the in-plane magnetic field $\mathbf{B}_{\parallel}(t) = B_1(r)(-\sin(\omega t + m\varphi), \cos(\omega t + m\varphi))$ rotates either in a counterclockwise or clockwise fashion, representing the left or right circularly polarized light beam. The additional integer-valued quantum number m in (7.44) refers to the orbital angular momentum of light. Solving Maxwell's equation yields the radial profile $B_1(r)$, which turns out to be a product of the Gaussian envelope and a modified Laguerre function, and the resulting beam structure is known as the Laguerre–Gaussian beam. Due to the phase singularity present for $m \neq 0$ modes, the optical beam is also known as the optical vortex. Numerical simulations of the LLG equation with the time-dependent Zeeman term (7.44) have demonstrated that nucleation of skyrmions is feasible with a properly tailored optical beam [11].

The physics of the circularly polarized beam can be best seen in the rotating frame. First we write $(n^x, n^y) = (\sin\theta\cos\psi, \sin\theta\sin\psi)$, using the symbol ψ for the azimuthal angle to avoid potential confusion with the angle φ used to describe the in-plane magnetic field. The in-plane part of the Zeeman term becomes

$$-B_0(r)\sin\theta\sin(\psi - \omega t - m\varphi). \qquad (7.45)$$

Next, suppose we move to a rotating frame of reference $\psi \to \psi + \omega t$, and simultaneously execute the same O(2) rotation of the space coordinates, $\mathbf{r} \to \mathcal{R}(\omega t)\mathbf{r}$. After these transformations, the Zeeman Hamiltonian becomes time-independent,

$$H_Z = -B_0 n^z - B_1(r)\sin(\psi - m\varphi), \qquad (7.46)$$

while the rest of the Hamiltonian will remain invariant under the simultaneous rotation of spin and space. The best configuration to take advantage of the in-plane Zeeman energy is achieved if the "swirl" of the in-plane spin component (n^x, n^y) matches that of the in-plane magnetic field. For the anti-skyrmion solution we have been discussing with $D > 0$, this would be consistent with taking $\psi = \varphi + \pi/2$, to match the $m = +1$ pattern of the in-plane magnetic field, or the $m = +1$ optical vortex.

The time-dependent transformation $\psi \to \psi + \omega t$ also modifies the Berry phase part of the Lagrangian,

$$- S(1 - \cos\theta)(\partial_t \psi + \omega), \tag{7.47}$$

effectively altering the out-of-plane Zeeman field to $B_0 + S\omega$ for those spatial regions where the circularly polarized beam is focused. If the frequency of light is chosen to match the typical exchange energy $\hbar\omega \sim J$, the effective Zeeman field can easily overpower the external field B_0, estimated to be of order $D^2/J \ll J$.

We are reminded, however, that actual skyrmion nucleation by an intensely focused (in both spatial and time domains) optical beam is complicated by heating effects, generation and subsequent emission pf spin waves, and so forth. Another technical challenge to be overcome is the ability to focus the beam onto an area as small as the size of a skyrmion.

7.3 Skyrmion Nucleation and Tunneling

7.3.1 *A Brief Theory of Nucleation*

One can discuss the skyrmion nucleation process within the traditional framework of instanton calculus. Even in a thermodynamic environment where the skyrmions are not expected to be the stable objects, a local variation of the magnetic field could take place favoring the skyrmion formation within a small region. Spin-polarized STM experiments performed on a PdFe bilayer deposited on an Ir(111) substrate proved to be a useful case in point [12]. In this example, the magnetic bilayer experiences a transition from the spiral to the skyrmion crystal phase when the perpendicular magnetic field is increased, as in other thin-film chiral magnets. In this system, the origin of the DM interaction is likely the breaking of inversion symmetry at the Ir substrate and PdFe bilayer interface. In any case, at magnitudes of the external field that were too weak to sustain the skyrmion lattice phase, a local injection of spin-polarized current was able to nucleate an individual skyrmion below the STM tip. An experiment such as this one raises the hope of controlled "on" and "off" switching of skyrmion states by electrical means, which would be a key step in any attempt to build electronic circuitry out of skyrmion components.

Having the interesting potential raised by such STM experiments in mind, we attempt to develop a simple model of the skyrmion nucleation process. First of all, we note that skyrmion nucleation can be visualized as a process in which the radius R of the skyrmion evolves continuously from zero to the equilibrium value R_s. Taking the model anti-skyrmion configuration we have used throughout this work,

$$\mathbf{n}_s = \left(-\frac{2Ry}{r^2 + R^2}, \frac{2Rx}{r^2 + R^2}, \frac{r^2 - R^2}{r^2 + R^2} \right), \tag{7.48}$$

we see that as R shrinks to zero the spin configuration becomes ferromagnetic, $\mathbf{n} = (0, 0, 1)$. The problem of calculating the nucleation rate for skyrmion can then

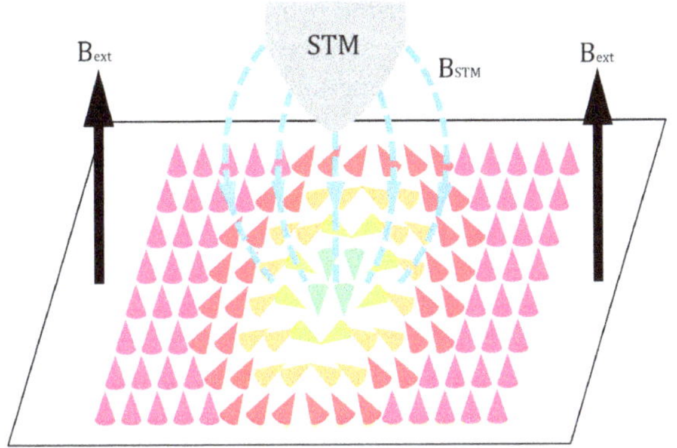

Fig. 7.2 Schematic view of the local environment change that could lead to a skyrmion nucleation in the film. While the external field is strong enough to polarize the spins completely, a local variation of the field due to the nearby STM tip can reduce the magnetic field to a level where a skyrmion can nucleate

be replaced by one of describing the evolution of a single variable $R(\tau)$ from 0 to R_s in imaginary time. Following the standard instanton technique, one can write down the imaginary time action for $R(\tau)$ and solve the saddle-point equation to obtain the nucleation rate. In this approach, the entire nucleation process is contained in the dynamics of a single collective variable $R(\tau)$. Previously, we used another kind of collective coordinate approach in which the skyrmion's center $\mathbf{R}(t)$ was treated as the sole dynamical variable. In the nucleation process considered here, we fix the center position and only the skyrmion radius $R(\tau)$ is considered to have the (imaginary) time dependence (Fig. 7.2).

In the CP^1 representation, the temporal part of the spin Lagrangian is written as $-2S\int d^2\mathbf{r}d\tau\, a_0(\tau)$, where $a_0(\tau) = -i\mathbf{z}^\dagger\partial_\tau\mathbf{z}$ [cf. (1.65)]. The skyrmion spin configuration (7.48) is tantamount to the CP^1 configuration

$$\mathbf{z}_s(\tau) = \frac{1}{\sqrt{r^2 + R(\tau)^2}}\begin{pmatrix} x - iy \\ iR(\tau)\end{pmatrix}, \quad \text{(first try)} \tag{7.49}$$

where the time dependence of the skyrmion radius $R(\tau)$ is written out explicitly. It turns out, however, that the geometric phase that follows from $\mathbf{z}_s(\tau)$ is zero, i.e., $\mathbf{z}_s^\dagger\partial_\tau\mathbf{z}_s = 0$, and it looks as though the whole instanton scheme breaks down from the start.

Fortunately, we can use a mathematical trick to get around this difficulty by allowing the function $R(\tau)$ to become complex-valued,

$$\mathbf{z}_s(\tau) = \frac{1}{\sqrt{r^2 + |R(\tau)|^2}}\begin{pmatrix} x - iy \\ iR(\tau)\end{pmatrix}. \quad \text{(second try)} \tag{7.50}$$

With this modification, we obtain the gauge field component

$$a_0(\tau) = \frac{i}{2}\frac{1}{r^2 + |R(t)|^2}\left(R\frac{dR^*}{d\tau} - R^*\frac{dR}{d\tau}\right). \tag{7.51}$$

Integrating $a_0(\tau)$ over the two-dimensional space yields a logarithmic divergence, similar to what we encountered several times in the variational energy calculation, which was due to the long-tail character of $\mathbf{n}_s$. Assuming a reasonable cutoff value, the space integral in the CP^1 action gives

$$\int d^2\mathbf{r}\, a_0(\tau) = \frac{i}{4} h\rho(R\dot{R}^* - R^*\dot{R}) = \frac{1}{2} h\rho(X\dot{Y} - Y\dot{X}), \tag{7.52}$$

where an unknown factor with the dimension of inverse (length)^2 is summarized as ρ, and the complex radius R has been audaciously replaced by $R = X + iY$. This exercise shows that the temporal part of the instanton action becomes identical to the Berry phase part of the vortex or skyrmion action. The problem of skyrmion nucleation is then equivalent to calculating the probability of a topological point-particle object tunneling in some potential barrier $V(X, Y)$, subject to the boundary conditions $\mathbf{R} = (0, 0)$ at $\tau = 0$ and $\mathbf{R} = (R_s, 0)$ at $\tau = \infty$. The potential function $V(X, Y)$ follows from substitution of the complexified CP^1 wave function (7.50) into the Hamiltonian. A thorough calculation of the skyrmion nucleation rate can be found in Ref. [13].

7.3.2 A Brief Theory of Tunneling

Next, we move on to exploring the possibility of skyrmion tunneling. Assuming the pre-existence of the skyrmion on one side of the barrier in a linear strip, such as shown in Fig. 7.3, we consider the probability of the skyrmion tunneling through the barrier to emerge on the other side while driven by the electronic current along the strip.

In Chap. 5, we derived the following (dissipationless) skyrmion action that includes the contributions arising from the coupling to the electronic current [cf. (5.3.2)]:

$$L_{s,e} = \frac{1}{2} M_s \dot{\mathbf{R}}^2 - \frac{1}{2}[G + hJ_0^e]\hat{z} \cdot (\mathbf{R} \times \dot{\mathbf{R}}) + h\hat{z} \cdot (\mathbf{R} \times \mathbf{J}^e) - V(\mathbf{R}). \tag{7.53}$$

For convenience, we abbreviate $G + hJ_0^e$ as G from here on. The simplest tunneling problem would be the one taking place in one dimension, but in one-dimensional systems the Berry phase effect vanishes. The next simplest geometry where interesting effects are expected is the strip geometry, assuming a finite width in the y-direction

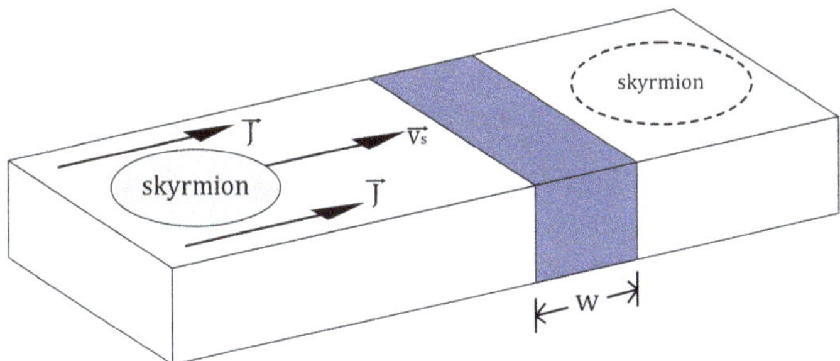

Fig. 7.3 A possible tunneling geometry for skyrmions driven by electronic current. A tunneling barrier of width W is inserted in a narrow bar, and an electric current drives the skyrmion forward. If the current is too weak, or the potential barrier is sufficiently high, the skyrmion can only emerge on the other side by tunneling through the barrier

and an infinite extent in the x-direction. Without much loss of generality, a potential barrier of the simple form

$$\begin{aligned} V(X,Y) &= V_0\left[1+\left(\frac{Y}{L}\right)^2\right] \quad (0<X<W), \\ &= 0 \qquad\qquad\qquad\qquad \text{(otherwise)} \end{aligned} \tag{7.54}$$

can be assumed, where the constant V_0 represents the potential raised within the tunneling region against the rest of the space. The width of the constriction in the Y-direction is denoted by L.

The real-time equation of motion within the potential barrier region ($0 < X < W$) is given by

$$\begin{aligned} M_s\ddot{X} + G\dot{Y} &= 0 \\ M_s\ddot{Y} - G\dot{X} + J + \frac{2V_0}{L^2}Y &= 0. \end{aligned} \tag{7.55}$$

Here, we consider the current J to be uniform throughout the whole strip, including the tunneling region. Assuming that the skyrmion is passing through the barrier region by means of the tunneling process, one can rewrite (7.55) in terms of the imaginary time $t \to -i\tau$, after completing the coordinate shift $Y \to Y + Y_0$, $Y_0 = -JL^2/2V_0$:

$$\begin{aligned} M_s\ddot{X} - iG\dot{Y} &= 0 \\ M_s\ddot{Y} + iG\dot{X} &= \frac{2V_0}{L^2}Y. \end{aligned} \tag{7.56}$$

The boundary conditions on the skyrmion coordinates are

$$(X(0),Y(0)) = (0,0), \quad (X(T),Y(T)) = (W,0). \tag{7.57}$$

In the massless limit, $M_s = 0$, a solution that meets the boundary conditions (7.57) is easy to find. From the first equation we have $Y(\tau) = 0$, and the second equation $\dot{X} = 0$ together with the boundary condition is solved for $X(\tau) = (W/T)\tau$. Inserting such solution into the massless skyrmion action gives

$$L_{s,e} = -JY_0 - V_0\left[1 + \left(\frac{Y_0}{L}\right)^2\right] = \frac{L^2J^2}{4V_0} - V_0. \tag{7.58}$$

The resulting tunneling probability is expressed as

$$(\text{Prob.}) \sim \exp\left(-2T\left[V_0 - \frac{L^2J^2}{4V_0}\right]\right). \tag{7.59}$$

Since the tunneling probability cannot exceed one, (7.59) implies that the tunneling regime is confined to

$$J \leq 2V_0/L. \tag{7.60}$$

In order to ensure that we observe a true tunneling event, the potential V_0 in the tunneling region must be raised high enough so that classical trajectories are forbidden. Interestingly, we see that the tunneling condition is controlled by the magnitude of the current J. Thus, for skyrmions entering the tunneling region with a sufficiently big impulse from the current, the effective tunneling barrier can be made low enough to allow classical passage.

Tunneling rates for the massive skyrmion can be found by numerical calculation. The first integral of motion can be derived from (7.56),

$$E = -\frac{1}{2}M_s(\dot{X}^2(\tau) + \dot{Y}^2(\tau)) + \frac{V_0}{L^2}Y^2(\tau). \tag{7.61}$$

The total energy E is then fixed by the initial condition of the skyrmion entering the tunneling region, while the tunneling time T enters the solution $(X(\tau), Y(\tau))$ implicitly, through the need to meet the boundary conditions (7.57).[4] The current-driven skyrmion tunneling problem presented here finds an obvious analogue in vortex tunneling driven by a supercurrent, as studied in Ref. [14].

7.4 Skyrmion Dynamics in Multiferroic Insulators

Generally speaking, skyrmions belonged to the proprietary realm of metallic chiral magnets until they were discovered in an insulating multiferroic material Cu_2OSeO_3. Despite its vastly different electrical properties, the Cu_2OSeO_3 phase diagram proved

[4]The tunneling calculations presented here were done in collaboration with Sang-Jin Lee and Eun-Gook Moon.

to be exceedingly similar to those of metallic chiral magnets, pointing to the generality of the skyrmion phase in noncentrosymmetric ferromagnets.

With the absence of conduction electrons in an insulator, however, one loses the ability to control skyrmions by electrical means. Fortunately, magnetic ordering in Cu_2OSeO_3 is accompanied by the formation of electric dipole moments due to the multiferroic nature of the material. The mechanism for the dipole moment formation in Cu_2OSeO_3 is known to be the *p*-*d* hybridization mechanism of multiferroicity. What this means is summed up in the following very simple formula [15, 16]:

$$\mathbf{P}_i = \lambda(S_i^y S_i^z, S_i^z S_i^x, S_i^x S_i^y). \tag{7.62}$$

The site i here represents one unit cell of the Cu_2OSeO_3 crystal, and λ is a measure of the dipole moment $\mathbf{P}_i$ induced at the same unit cell by the magnetic moment $\mathbf{S}_i$. The magneto-electric Hamiltonian is written in the usual way,

$$H_{\rm ME} = -\sum_i \mathbf{P}_i \cdot \mathbf{E}_i = -\lambda \sum_i (E_i^x S_i^y S_i^z + E_i^y S_i^z S_i^x + E_i^z S_i^x S_i^y). \tag{7.63}$$

Calculating the induced dipole moment $\mathbf{P}_i$ for a given skyrmion spin configuration, one discovers that the net dipole moment vanishes for skyrmions formed in the xy plane, or for magnetic fields applied parallel to the [001] crystal direction, but it points along the [111] direction if the magnetic field orientation is [111]. Choosing this particular field orientation so as to induce a finite dipole moment, one can start characterizing the skyrmion both by its topological charge Q_s and its net dipole moment P. One can also maximize the influence of the electric field coupling by orienting the $\mathbf{E}$-field parallel to the dipole moment, $\mathbf{P}_i \cdot \mathbf{E}_i \rightarrow P_i E_i$. Taking the continuum limit of the magneto-electric Hamiltonian then gives

$$H_{\rm ME} = -\lambda \int d^2\mathbf{r}\, \rho_D(\mathbf{r}) E(\mathbf{r}), \tag{7.64}$$

with a suitable redefinition of the coupling constant λ. Treating each skyrmion as a point-particle object allows us to write the dipole moment density

$$\rho_D(\mathbf{r}) = \sum_i p\delta^2(\mathbf{r} - \mathbf{r}_i), \tag{7.65}$$

where p represents the integrated dipole moment per skyrmion. Moreover, the magneto-electric Hamiltonian (7.64) becomes

$$H_{\rm ME} = -\lambda p \sum_i E(\mathbf{r}_i). \tag{7.66}$$

This expression is in complete analogy with the electric potential of a system of charges $q_e \sum_i V(\mathbf{r}_i)$.

In further analogy with the electrostatic force, one expects that the skyrmions are acted on by a force

$$\mathbf{F}_{\mathrm{ME}} = -\nabla H_{\mathrm{ME}}, \tag{7.67}$$

originating from the spatially modulated electric field. As a result, drift motion of skyrmions in a perpendicular direction to $\mathbf{F}_{\mathrm{ME}}$ will take place. Furthermore, the magneto-electric coupling can be used to excite the skyrmion with an oscillating electric field. Numerical Integration of the LLG equation with HDMZ Hamiltonian and including the magneto-electric Hamiltonian has shown the drift motion of skyrmions under a graded electric field and has further demonstrated the existence of microwave-accessible low-energy excitations in the skyrmion crystal that can be induced by an ac electric field [16].

References

1. Schütte, C., Garst, M.: Magnon-skyrmion scattering in chiral magnets. Phys. Rev. B **90**, 094423 (2014)
2. Sampaio, J., Cros, V., Rohart, S., Thiaville, A., Fert, A.: Nucleation, stability and current-induced motion of isolated magnetic skyrmions in nanostructures. Nat. Nanotech. **8**, 839 (2013)
3. Zhou, Y., Iacocca, E., Awad, A.A., Dumas, R.K., Zhang, F.C., Braun, H.B., Åkerman, J.: Dynamically stabilized magnetic skyrmions. Nat. Commun. **6**, 8193 (2015). doi:10.1038/ncomms9193
4. Yin, G., Li, Y., Kong, L., Lake, R.K., Chien, C.L., Zang, J.: Topological charge analysis of ultrafast single skyrmion creation. Phys. Rev. B **93**, 174403 (2016)
5. Yuan, H.Y., Wang, X.R.: Skyrmion creation and manipulation by nano-second current pulses. Sci. Rep. **6**, 22638 (2016). doi:10.1038/srep22638
6. Tchoe, Y., Han, J.H.: Skyrmion generation by current. Phys. Rev. B **85**, 174416 (2012)
7. Iwasaki, J., Mochizuki, M., Nagaosa, N.: Current-induced skyrmion dynamics in constricted geometries. Nat. Nanotechnol. **8**, 742 (2013). doi:10.1038/nnano.2013.176
8. Jiang, W., Upadhyaya, P., Zhang, W., Yu, G., Jungfleisch, M.B., Fradkin, F.Y., Pearson, J.E., Tserkovnyak, Y., Wang, K.L., Heinonen, O., te Velthuis, S.G.E., Hoffmann, A.: Blowing magnetic skyrmion bubbles. Science **349**, 283 (2015)
9. Woo, S., Litzius, K., Krüger, B., Im, M.Y., Caretta, L., Richter, K., Mann, M., Krone, A., Reeve, R.M., Weigand, M., Agrawal, P., Lemesh, I., Mawass, M.A., Fischer, P., Kläui, M., Beach, G.S.D.: Observation of room-temperature magnetic skyrmions and their current-driven dynamics in ultrathin metallic ferromagnets. Nat. Mater. **15**, 501 (2016). doi:10.1038/nmat4593
10. Yu, X., Morikawa, D., Tokunaga, Y., Kubota, M., Kurumaji, T., Oike, H., Nakamura, M., Kagawa, F., Taguchi, Y., Arima, T., Kawasaki, M., Tokura, Y.: Current-induced nucleation and annihilation of magnetic skyrmions at room temperature in a chiral magnet. Adv. Mater. **29**, 1606178 (2017). doi:10.1002/adma.201606178
11. Fujita, H., Sato, M.: Encoding orbital angular momentum of lights in magnets. arXiv:1612.00176 (2016)
12. Romming, N., Hanneken, C., Menzel, M., Bickel, J.E., Wolter, B., Bergmann, K.V., Kubetzka, A., Wiesendanger, R.: Writing and deleting single magnetic skyrmions. Science **341**, 636 (2013)

13. Diaz, S.A., Arovas, D.P.: Quantum nucleation of skyrmions in magnetic films by inhomogeneous fields. arXiv:1604.0401 (2016)
14. Ao, P., Thouless, D.J.: Tunneling of a quantized vortex: roles of pinning and dissipation. Phys. Rev. Lett. **72**, 132 (1994)
15. Seki, S., Ishiwata, S., Tokura, Y.: Magnetoelectric nature of skyrmions in a chiral magnetic insulator Cu_2OSeO_3. Phys. Rev. B **86**, 060403(R) (2012)
16. Liu, Y.H., Li, Y.Q., Han, J.H.: Skyrmion dynamics in multiferroic insulators. Phys. Rev. B **87**, 100402(R) (2013)

Chapter 8
Skyrmions in Spinor Bose-Einstein Condensates

As a physical medium, cold atoms are wildly different from solid-state magnetic materials. Yet, some aspects of cold atom physics are subject to the same homotopy analysis we performed in Chap. 2, and relevant topological excitations can be classified accordingly. Indeed, vortices, skyrmions, and other interesting exotic topological excitations are possible states in spinor condensates, which are the quantum gases made up of atoms with spin. In contrast to the solid-state materials, however, the size of the effective spin in such spinor condensates can become very large for some atoms, complicating the topological analysis. In this chapter, we discuss various topological excitations for the two simplest cases of spinor condensates made of two- and three-component spins.

8.1 Topological Excitations of Two-Component Spinors

A classic superfluid such as ^{4}He is described by a complex scalar order parameter

$$\Psi = \sqrt{\rho}\psi, \tag{8.1}$$

where, after taking out the density part $\sqrt{\rho}$, ψ becomes a unit modulus complex scalar $\psi^*\psi = 1$, which is simply the phase angle $e^{i\theta}$ of the superfluid. The Hamiltonian of such a scalar superfluid consists of the kinetic energy and the contact interaction ($\hbar \equiv m \equiv 1$):

$$H = H_{kin} + H_{int} = \frac{1}{2}\nabla\Psi^* \cdot \nabla\Psi + \frac{g_0}{2}(\rho - \rho_0)^2. \tag{8.2}$$

If we consider the case in which the interaction energy far outweighs the kinetic part in strength, the physics of the condensate can be conveniently analyzed around the

J.H. Han, *Skyrmions in Condensed Matter*, Springer Tracts in Modern Physics 278, https://doi.org/10.1007/978-3-319-69246-3_8

potential energy minimum $\rho = \rho_0$. The interaction energy is "quenched" by taking $\rho = \rho_0$, leaving the freedom to choose the phase part, $\psi = e^{i\theta}$, without changing the ground state energy. One says that the ground state manifold has the U(1) symmetry, as it consists of all possible phase angles of the order parameter Ψ. Whether there are interesting topological excitations allowed in such manifold of ground states can be answered by examining the homotopy of the U(1) manifold:

$$\pi_1(U(1)) = \mathbb{Z}, \quad \pi_2(U(1)) = \pi_3(U(1)) = 0. \tag{8.3}$$

The nontrivial π_1 homotopy corresponds to the creation of vortices of integer topological charge, while the creation of skyrmions is ruled out by the trivial second homotopy $\pi_2(U(1)) = 0$. This is why one speaks only of vortex excitations in traditional scalar superfluid ^{4}He.

We can classify the ground-state manifold of the more complex order parameter field describing the condensate of spinor bosons in a similar fashion. First, the two-component spinor condensate is described by the wave function

$$\Psi = \sqrt{\rho}\psi = \sqrt{\rho}\begin{pmatrix} \psi_1 \\ \psi_2 \end{pmatrix}, \tag{8.4}$$

where the uni-modular constraint $|\psi_1|^2 + |\psi_2|^2 = \psi^\dagger\psi = 1$ casts each spinor ψ as a point on the three-sphere S^3. Alternatively, one can factor out the overall phase and write $\psi = e^{i\theta}\mathbf{z}$, writing the spinor as a product of the U(1) phase and the CP^1 field $\mathbf{z}$. Clearly, the two manifolds are equivalent,

$$S^3 \cong \mathrm{U}(1) \times \mathrm{CP}^1. \tag{8.5}$$

The Hamiltonian for the two-component cold atom system is

$$H = \frac{1}{2}\int (\nabla\Psi^\dagger)\cdot(\nabla\Psi) + H_{int}. \tag{8.6}$$

Typically, the interaction part of the spinor Hamiltonian H_{int} considers the short-range collisions of the two boson particles through their density $\Psi^\dagger\Psi = \rho$ and their spin $\Psi^\dagger\boldsymbol{\sigma}\Psi$ as

$$H_{int} = \frac{g_0}{2}(\Psi^\dagger\Psi - \rho_0)^2 + \frac{g_2}{2}(\Psi^\dagger\boldsymbol{\sigma}\Psi)^2. \tag{8.7}$$

The strength of each interaction is characterized by the constants g_0 and g_2 and, experimentally, we find $g_0 \gg |g_2|$. In order to identify the proper ground-state topology we take $g_2 = 0$ and quench the kinetic energy, leaving the ground state manifold to be $\rho = \rho_0$. It follows that the ground state manifold is precisely that of the uni-modular spinor ψ in (8.4), i.e., S^3. Our discussion of the possible topological excitations in the spin-1/2 condensate then proceeds by identifying the various homotopies of S^3, which are

$$\pi_1(S^3) = \pi_2(S^3) = 0, \quad \pi_3(S^3) = \mathbb{Z}. \tag{8.8}$$

The first two trivial homotopies indicate the impossibility of constructing a stable vortex or a two-dimensional baby skyrmion in a spin-1/2 condensate. Instead, the third homotopy indicates that it is possible to construct a topologically stable three-dimensional skyrmion solution. An extensive discussion on the mathematical construction of the three-dimensional skyrmion was provided earlier in Sect. 2.5.

When a skyrmion is created, there is a corresponding kinetic energy cost due to the spatially varying texture of the wave function. Up until now, we have ignored this cost as minute, by assumption, compared to the interaction energy, at least in the region far away from the singularity. A good way to think about the kinetic energy cost is to restrict the Hilbert space entirely to those states allowed in the $g_0 \to \infty$ limit. The interaction part of the Hamiltonian can then be removed, leaving only the kinetic energy term. Expressed within the reduced Hilbert space $\Psi = \sqrt{\rho_0}\psi$, we have:

$$H_{kin} = \frac{1}{2}\rho_0(\nabla\psi^\dagger \cdot \nabla\psi) = \frac{1}{2}\rho_0 \sum_{\mu=1}^{d} \partial_\mu \mathbf{N} \cdot \partial_\mu \mathbf{N}. \tag{8.9}$$

The order parameter $\mathbf{N} = (N_1, N_2, N_3, N_4)$ defines a unit vector on the three-sphere S^3. The spatial components in the above expression span $\mu = 1$ through d.

The O(4) nonlinear σ-model written above takes on a more physical form when the uni-modulus spinor ψ is rewritten in the $\mathrm{U}(1) \times \mathrm{CP}^1$ form. Writing $\psi = e^{i\theta}\mathbf{z}$, one can show

$$\sum_\mu \partial_\mu \mathbf{N} \cdot \partial_\mu \mathbf{N} = \frac{1}{4}\sum_\mu \partial_\mu \mathbf{n} \cdot \partial_\mu \mathbf{n} + \frac{1}{2}(\nabla\theta + \mathbf{a})^2, \tag{8.10}$$

where the vector potential $\mathbf{a} = -i\mathbf{z}^\dagger \nabla \mathbf{z}$ is the familiar gauge field derived from the CP^1 field and the first term on the right-hand side is the familiar O(3) nonlinear σ-model constructed from the local magnetization $\mathbf{n}$. The local superfluid velocity $\mathbf{v}$ is given by the gauge-invariant combination (restoring $\hbar/m$ momentarily)

$$\mathbf{v} = \frac{\hbar}{m}(\nabla\theta + \mathbf{a}). \tag{8.11}$$

This mapping makes it clear that the kinetic energy of the spin-1/2 condensate separates into a "magnetic part" $(\partial_\mu \mathbf{n} \cdot \partial_\mu \mathbf{n})$ and a "mass part" $(\mathbf{v}^2)$. The true order parameter is still embodied in the O(4) vector $\mathbf{N}$, and the relevant topological excitation is the one realized in three space dimensionality. An explicit construction of this topological state can be found in Ref. [1].

The g_2-dependent energy is a function of the local magnetization vector $\psi^\dagger \boldsymbol{\sigma} \psi = \mathbf{m}$. As $|\mathbf{m}| = 1$ regardless of the orientation of the vector $\mathbf{m}$, the order parameter space is not narrowed down any further when $g_2 \neq 0$. In fact, in the particular case of

$S = 1/2$ the g_2-term simply drops out as constant. To the knowledgeable reader, our discussion of the null influence of the g_2-term may have seemed obvious. The actual purpose of going through this exercise was to prepare the reader for the discussion of the topology of the three-component $S = 1$ cold atom system in the following section. There, the classification of possible order parameters becomes far more intricate, as does the associated problem of identifying the possible topological excitations.

The hydrodynamic form of the action for the two-component spinor condensate can be obtained by writing

$$\Psi = \sqrt{\rho} e^{i\theta_s} e^{i\theta_v} \mathbf{z}, \tag{8.12}$$

where the phase angle has been separated into its smooth and vortex parts, as in the one-component theory. The hydrodynamic Lagrangian for the two-component spinor is then[1]

$$\begin{aligned} L = & -\rho(\partial_t \theta_s + a_t - i\chi_v^* \partial_t \chi_v) - \frac{\rho_0}{2m}(\nabla\theta_s + \mathbf{a} - i\chi_v^* \nabla \chi_v)^2 \\ & - \frac{\rho_0}{4m}(\partial_\mu \mathbf{n})^2 - \frac{(\nabla\rho)^2}{8m\rho} - \frac{g_0}{2}(\rho - \rho_0)^2, \end{aligned} \tag{8.13}$$

where the vortex phase field has been abbreviated by $\chi_v = e^{i\theta_v}$. Note that the emergent gauge field $a_\mu = -i\mathbf{z}^\dagger \partial_\mu \mathbf{z}$ appears alongside the U(1) gauge field $-i\chi_v^* \partial_\mu \chi_v$ in the Lagrangian. In fact, the first line of the hydrodynamic action is quite similar to that which was obtained in Sect. 4.1.3 [cf. (4.64)] for the one-component condensate, the sole difference being the addition of the gauge field **a** from the CP[1] field. The duality transformation we used for the one-component theory works here with very little modification, resulting in the Lagrangian

$$L = \frac{1}{2} \cdot \frac{2m}{\rho_0} \mathbf{J}^2 - \frac{(\nabla J_0)^2}{8m\rho_0} - \frac{\rho_0}{4m}(\partial_\mu \mathbf{n})^2 - \frac{g}{2}(J_0 - \rho_0)^2 - c_\alpha J_\alpha^t, \tag{8.14}$$

where the index $\alpha = t, x, y$ runs over $(2+1)$-dimensional spacetime.

The topological three-current J_α^t consists of two pieces:

$$\begin{aligned} J_\alpha^t &= J_\alpha^v + J_\alpha^s \\ J_\alpha^v &= \frac{1}{2\pi} \varepsilon_{\alpha\beta\gamma} \partial_\beta [-i\chi_v^* \partial_\gamma \chi_v] \\ J_\alpha^s &= \frac{1}{2\pi} \varepsilon_{\alpha\beta\gamma} \partial_\beta a_\gamma . \end{aligned} \tag{8.15}$$

The first one J_α^v is the vortex current familiar from the duality of the U(1) theory. The second term, J_α^s, has also been seen before in (2.72), and is the CP[1] representation of

[1]Our derivation of the hydrodynamic action through a duality transformation follows from that of Kane and Lee [2].

the topological (skyrmion) current in (2 + 1)-dimensions. It should be emphasized, however, that we do not have separate conservation laws for the vortex current or the skyrmion current; only their sum obeys the conservation law.

Readers are also reminded of the definition of the three-current[2]:

$$J_\alpha = (\rho, \mathbf{J}) = \frac{1}{2\pi}\varepsilon_{\alpha\beta\gamma}\partial_\beta c_\gamma, \tag{8.16}$$

where the two-dimensional vector **J** is the Hubbard–Stratonovich field. The interaction constraint $\rho = \rho_0$ restricts the temporal component to obey

$$J_t = \frac{1}{2\pi}\nabla_2 \times \mathbf{c} = \rho_0, \tag{8.17}$$

which is solved by $\mathbf{c} = \pi\rho_0(-y, x)$. Thus, vortices and skyrmions scattered around the two-dimensional space are described by the topological three-current

$$J^t_\alpha = \sum_v q_v(1, -\dot{X}_v, -\dot{Y}_v)\delta^2(\mathbf{r} - \mathbf{R}_v) + \sum_s Q_s(1, -\dot{X}_s, -\dot{Y}_s)\delta^2(\mathbf{r} - \mathbf{R}_s). \tag{8.18}$$

The spatial component of the coupling $-c_\alpha J^t_\alpha$ in (8.14) is therefore reduced to the geometric action

$$\frac{1}{2}h\rho_0\sum_v q_v\hat{z}\cdot(\mathbf{R}_v \times \dot{\mathbf{R}}_v) + \frac{1}{2}h\rho_0\sum_s Q_s\hat{z}\cdot(\mathbf{R}_s \times \dot{\mathbf{R}}_s). \tag{8.19}$$

Despite the apparent separation of the topological degrees of freedom into vortex and skyrmion parts, there is no genuine meaning to the vortex or skyrmion excitations *per se* in this theory, because they fail to obey separate conservation laws. As dictated by the homotopy, the only stable topological object in the two-component spinor is found in three-dimensional space.

8.2 Topology of Three-Component Spinors and Solitons

In the one-component case, the only stable excitations that are permitted are vortex excitations. In two-component spinor gases, strictly speaking, neither vortex nor two-dimensional skyrmion was allowed, and only three-dimensional topological defects was possible. We move now to a discussion of the three-component spinor state defined by the wavefunction

[2] In the U(1) theory we used a_μ instead of c_μ. Now, since a_μ is used to express the CP^1 gauge field, and b_μ to express the emergent flux, the next available symbol is c_μ.

$$\Psi = \sqrt{\rho}\psi = \sqrt{\rho}\begin{pmatrix}\psi_1\\ \psi_2\\ \psi_3\end{pmatrix}. \tag{8.20}$$

The uni-modulus spinor ψ defines a point on the five-sphere S^5. Equivalently, by rewriting $\psi = e^{i\theta}\mathbf{z}$, the spinor can be viewed as a U(1) phase times the CP^2 field. By necessity, we have the equivalence

$$S^5 \cong \text{U}(1) \times \text{CP}^2. \tag{8.21}$$

As was noted in (8.7), the interaction part of the Hamiltonian in these systems consists of density-dependent and spin-dependent parts, with the density-dependent part dominating over the spin part. As before, we begin our classification of allowed order parameters by first seeking to minimize the density-dependent interaction. By assuming a uniform density $\rho = \rho_0$, the residual spinor part ψ is free to choose any value on the S^5 sphere without changing the energy.

The homotopy group for S^5 is rather boring:

$$\pi_d(S^5) = 0, \ d = 1, 2, 3. \tag{8.22}$$

In short, the order parameter space S^5 is simply "too big" to be wrapped around by the packaging options one has in lower dimensionality, much like when one tries to wrap a sphere with a rubber band only to see the band slip away. One must then go down to the next level in the energy hierarchy involving the spin-dependent interaction to discover nontrivial topological excitations. Depending on the sign of the spin-dependent interaction g_2, the condensate is either antiferromagnetic or ferromagnetic. We begin with the antiferromagnetic case, following the classification of topological excitations according to the sign of g_2 as proposed in Ref. [3].

8.2.1 Antiferromagnetic Spinor Space

At constant density, the spin-dependent interaction for the three-component spinor takes on the form

$$H_{int.} = \frac{1}{2}g_2\rho_0^2(\mathbf{m})^2, \quad \mathbf{m} = \psi^\dagger\mathbf{F}\psi. \tag{8.23}$$

The spin-1 matrices $\mathbf{F} = (F_x, F_y, F_z)$ are written in the angular momentum basis as $F_z = \text{diag}(+1, 0, -1)$, and in terms of ladder operators

$$F^+ = F_x + iF_y = \begin{pmatrix}0 & \sqrt{2} & 0\\ 0 & 0 & \sqrt{2}\\ 0 & 0 & 0\end{pmatrix}, \quad F^- = F_x - iF_y = (F^+)^\dagger. \tag{8.24}$$

The local magnetization vector $\mathbf{m}$ occupies a subset of the overall manifold S^5. Depending on which atoms are being trapped and cooled in an experiment, one has either $g_2 > 0$ or $g_2 < 0$. The former system goes by various names such as polar, antiferromagnetic, or nematic; here we adopt "antiferromagnetic". In marked difference to the $S = 1/2$ condensate, the magnetization vector $\mathbf{m}$ no longer needs to have a unit magnitude. Ideally, the spin-dependent energy is minimized by having $\mathbf{m} = 0$ everywhere in the case of g_2 very large and positive. Since the original manifold S^5 is so large, a subspace which satisfies the condition of zero magnetization, $\mathbf{m} = 0$, does indeed exist.

In order to generate a $\mathbf{m} = 0$ space, one starts with a reference state, $\psi_{m_z=0} = (0\ 1\ 0)^T$, where only the $\langle F_z \rangle = m_z = 0$ component is occupied by bosons. Acting on this state with the most general Euler rotation operator gives

$$U(\alpha, \beta, \gamma)\psi_{m_z=0} = e^{-i\alpha F_z} e^{-i\beta F_y} e^{-i\gamma F_z} \psi_{m_z=0} = \begin{pmatrix} -\frac{1}{\sqrt{2}} e^{-i\alpha} \sin\beta \\ \cos\beta \\ \frac{1}{\sqrt{2}} e^{i\alpha} \sin\beta \end{pmatrix}. \quad (8.25)$$

The first rotation $e^{-i\gamma F_z}$ produces no change when acting upon $\psi_{m_z=0}$, leaving only two independent variables to specify the antiferromagnetic spinor. One can readily check $\psi^\dagger \mathbf{F} \psi = 0$ for wavefunctions of this type.

There is a neat way to visualize the antiferromagnetic spinor defined above. With the two Eulerian angles α and β one can construct a unit vector

$$\mathbf{d} = (\sin\beta\cos\alpha, \sin\beta\sin\alpha, \cos\beta) \in S^2. \quad (8.26)$$

In terms of $\mathbf{d}$, the most general antiferromagnetic spinor including the U(1) phase becomes

$$\psi_A[\theta, \mathbf{d}] = e^{i\theta} \begin{pmatrix} -\frac{1}{\sqrt{2}}(d_x - i d_y) \\ d_z \\ \frac{1}{\sqrt{2}}(d_x + i d_y) \end{pmatrix} \equiv e^{i\theta} \eta_A[\mathbf{d}]. \quad (8.27)$$

The space of antiferromagnetic spinor wavefunctions satisfying the constraint $\mathbf{m} = 0$ therefore appears to be $\mathrm{U}(1) \times S^2$, where the U(1) factor arises from the phase $e^{i\theta}$ in the wavefunction. However, a more poised look at ψ_A reveals that a simultaneous change of θ to $\theta + \pi$ and $\mathbf{d}$ with $-\mathbf{d}$ leads to the identical wavefunction:

$$\psi_A[\theta, \mathbf{d}] = \psi_A[\theta + \pi, -\mathbf{d}]. \quad (8.28)$$

Mathematically, when two points on the original order parameter space are identified as one element, we are talking about the coset space,

$$\mathcal{M}_{\mathrm{AFM}} = U(1) \times S^2/\mathbb{Z}_2, \quad (8.29)$$

which is the true manifold for the three-component antiferromagnetic state. The homotopy theory for this manifold reveals

$$\pi_d(\mathscr{M}_{\mathrm{AFM}}) = \mathbb{Z}, \quad d = 1, 2, 3, \tag{8.30}$$

corresponding to a topologically stable vortex, two-dimensional skyrmion, and three-dimensional skyrmion, respectively.

Our next task is to construct each topological object explicitly. While doing so, it must be kept in mind that the U(1) phase and the S^2 vector are intertwined due to the identification of $(\theta, \mathbf{d})$ and $(\theta + \pi, -\mathbf{d})$. For starters, the quantity measuring the vortex number can be constructed as usual from the velocity vector

$$\mathbf{v}_A = -i\psi_A^\dagger(\nabla\psi_A) = \nabla\theta. \tag{8.31}$$

The $\mathbf{d}$-dependent part of the antiferromagnetic wavefunction $\eta_A[\mathbf{d}]$ drops out from the velocity since $\boldsymbol{\eta}_A^\dagger \partial_\mu \boldsymbol{\eta}_A = \partial_\mu(\mathbf{d}^2)/2 = 0$. In a scalar superfluid, the integral of the velocity field around a closed contour is an integer times 2π due to the uniqueness of the wavefunction. One might have expected the same argument to apply to the $S = 1$ antiferromagnetic condensate, had it not been for the critical property that $(\theta, \mathbf{d})$ and $(\theta + \pi, -\mathbf{d})$ are really the same wavefunction.

An interesting possibility for the half-quantum vortex in the A-phase of ^{3}He (which is different from the A-phase of MnSi!) was raised by Salomaa and Volovik [4]. Translated into the context of $S = 1$ antiferromagnetic condensate, it implies the possibility of a wavefunction such as $\theta = \varphi/2$, and $\mathbf{d} = (\cos\varphi/2, \sin\varphi/2, 0)$, where the azimuthal angle $\varphi = \tan^{-1}(y/x)$ is used. Combined, the wavefunction

$$\psi = e^{i\varphi/2}\begin{pmatrix} -\frac{1}{\sqrt{2}}e^{-i\varphi/2} \\ 0 \\ \frac{1}{\sqrt{2}}e^{i\varphi/2} \end{pmatrix} = \begin{pmatrix} -\frac{1}{\sqrt{2}} \\ 0 \\ \frac{1}{\sqrt{2}}e^{i\varphi} \end{pmatrix} \tag{8.32}$$

is perfectly well-defined everywhere except at the origin. Pictorially, the π-mismatch of the phase θ is compensated by the disclination in the spin sector $\mathbf{d}$. A similar construction with $\mathbf{d} = (\cos\phi/2, -\sin\phi/2, 0)$ gives the vorticity in the upper component $m_z = +1$. In any case, one finds that the circulation is quantized in half-integer steps,

$$\frac{1}{2\pi}\oint \mathbf{v}_A \cdot d\mathbf{r} = \frac{n_A}{2}, \tag{8.33}$$

and not by an integer multiple as in the U(1) vortex.

The exact core profile of a half-quantum vortex whose asymptotic behavior is that of (8.32) depends on the interplay among various interactions in the Hamiltonian. There is a general discussion one can make regarding the structure of the vortex core based on the evolving energetics as the core region is approached, which proves to be exotic compared to the core physics of the scalar U(1) superfluid vortex.

Assuming that both g_0- and g_2-dependent interactions are quenched with $\rho = \rho_0$ and $\mathbf{m} = 0$, the kinetic energy of the three-component condensate becomes

$$H_{kin} = \frac{\rho_0}{2}\nabla\psi^\dagger \cdot \nabla\psi \Rightarrow \frac{\rho_0}{2}\mathbf{v}_A^2 + \frac{\rho_0}{2}\sum_\mu (\partial_\mu \mathbf{d})^2. \tag{8.34}$$

In the vortex configuration, the diverging velocity generates enough kinetic energy within the core region to surpass the spin-dependent interaction energy. When this occurs, the homotopy consideration based on the dominance of g_2-interaction can no longer be valid. Instead, there will be a region where the kinetic energy of the vortex is greater than that stored in the g_2-part, but is still much weaker than the g_0-part. In such a region, only the density constraint $\rho = \rho_0$ remains valid, while the order parameter space has to be enlarged from $\mathscr{M}_{\mathrm{AFM}}$ to that of S^5. The local magnetization $\mathbf{m}$ outside the $\mathscr{M}_{\mathrm{AFM}}$ manifold is no longer zero, so this intermediate region can have some of the diverging kinetic energy redistributed to the magnetic sector. The core region, as a result, becomes nonsingular, while still maintaining the constant density $\rho = \rho_0$. This is markedly different from the ordinary U(1) vortex where the absence of a manifold bigger than U(1) necessarily forces $\rho \to 0$ condition at the core to resolve the kinetic energy crisis. It is a general lesson that when the order parameter space is large enough, one can avoid the diverging energy catastrophe of the vortex without creating the density depletion at the core.

8.2.2 *Ferromagnetic Spinor Space*

As we go to the opposite limit of g_2 large and negative, the magnetization vector $\mathbf{m}$ is required to be as large as possible, resulting in the so-called ferromagnetic spinor state. The condition on $\mathbf{m}$ is naturally met by the coherent state, achieved by Euler rotation of the reference state $\psi_{m_z=+1} = (1\;0\;0)^T$:

$$\mathscr{U}(\alpha, \beta, \gamma)\psi_{m_z=+1} = e^{-i\gamma}\begin{pmatrix} e^{-i\alpha}\cos^2\frac{\beta}{2} \\ \frac{1}{\sqrt{2}}\sin\beta \\ e^{i\alpha}\sin^2\frac{\beta}{2} \end{pmatrix}. \tag{8.35}$$

Multiplying by the global phase $e^{i\theta}$ changes $e^{-i\gamma}$ to $e^{i(\theta-\gamma)}$, and the general wavefunction that belongs to the ferromagnetic manifold is

$$\psi_F = e^{i(\theta-\gamma)}\begin{pmatrix} e^{-i\alpha}\cos^2\frac{\beta}{2} \\ \frac{1}{\sqrt{2}}\sin\beta \\ e^{i\alpha}\sin^2\frac{\beta}{2} \end{pmatrix} \equiv e^{i(\theta-\gamma)}\eta_F[\mathbf{d}], \tag{8.36}$$

where the local magnetization $\psi_F^\dagger \mathbf{F}\psi_F = \mathbf{m} = \mathbf{d}$ coincides with the $\mathbf{d}$-vector itself. While the U(1) phase depends on the combination $\theta - \gamma$, one might as well have ignored θ altogether and taken the whole order parameter space consistent with the constraint $\mathbf{m} \cdot \mathbf{m} = 1$ to be completely specified by (8.35), without the additional U(1) phase. The ferromagnetic order parameter space is, therefore, the same as the space of Euler rotations, or SO(3):

$$\mathscr{M}_{\mathrm{FM}} = \mathrm{SO}(3). \tag{8.37}$$

The homotopy groups for this particular manifold are

$$\pi_1(\mathscr{M}_{\mathrm{FM}}) = \mathbb{Z}_2, \quad \pi_2(\mathscr{M}_{\mathrm{FM}}) = 0, \quad \pi_3(\mathscr{M}_{\mathrm{FM}}) = \mathbb{Z}. \tag{8.38}$$

The first of these formulas implies that the vortex solution in the ferromagnetic manifold has only one possible topological number, +1. The other possibility is the state with topological number 0, topologically equivalent to a uniform solution. The $\mathbb{Z}_2$ character of the vortex is hard to grasp intuitively at first sight. Fortunately, one can understand the origin of the $\mathbb{Z}_2$ character of the topological solution by an explicit construction.

First, one needs a way of measuring these topological numbers. As with the antiferromagnetic manifold, the usual suspect is the velocity field $\mathbf{v}_F = -i\psi_F^\dagger \nabla \psi_F$,

$$\mathbf{v}_F = \nabla\theta + 2\mathbf{a}. \tag{8.39}$$

Here we are using θ instead of $\theta - \gamma$ for brevity. The vector potential $\mathbf{a}$ is the same as was defined earlier for the spin-1/2 condensate: $\mathbf{a} = -(\cos\beta\nabla\alpha)/2$, while the factor 2 multiplying $\mathbf{a}$ in the above formula marks the difference between the spin-1/2 and spin-1 coherent states. Due the uniqueness of the wavefunction, the integral of $\nabla\theta$ ought to be an integer multiple of 2π, which is the topological number n_F associated with the ferromagnetic vortex. On the other hand, this integer is related to the circulation and the emergent flux through

$$n_F = \frac{1}{2\pi}\oint \mathbf{v}_F \cdot d\mathbf{r} - \frac{1}{\pi}\oint (\nabla \times \mathbf{a}) \cdot d\mathbf{S}, \tag{8.40}$$

for an arbitrary loop encompassing the vortex core. It should be borne in mind that neither of the two integrals on the right-hand side has any reason to be integer-valued; only the difference has to match n_F. For two-dimensional space, one can further rewrite the last integral using the identity $\nabla_2 \times \mathbf{a} = (1/2)\mathbf{m} \cdot (\partial_x \mathbf{m} \times \partial_y \mathbf{m})$,

$$n_F = \frac{1}{2\pi}\oint \mathbf{v}_F \cdot d\mathbf{r} - \frac{1}{2\pi}\oint \mathbf{m} \cdot \left(\frac{\partial \mathbf{m}}{\partial x} \times \frac{\partial \mathbf{m}}{\partial y}\right) dxdy. \tag{8.41}$$

Now consider an extremely large loop practically encompassing the whole of the two-dimensional plane. The topological density on the far right-hand side, now integrated over the whole space, give rise to another integer, the well-known skyrmion number. This results in an alternative form of the above relation,

$$n_F + 2Q_s = \frac{1}{2\pi}\oint \mathbf{v}_F \cdot d\mathbf{r}. \tag{8.42}$$

The interpretation of this formula is as follows. A field configuration with, say, $n_F = 1$ and $Q_s = 0$, can be smoothly transformed into another configuration where n_F has decreased by 2 while Q_s has gone up by one, because the sum $n_F + 2Q_s$ representing the physical circulation remains the same in both cases. In general, configurations for which n_F differs by an even integer can be smoothly connected by the concomitant creation of a skyrmionic magnetization texture carrying some topological charge Q_s. It is still the case, however, that states sharing the same circulation $n_F + 2Q_s$ can have vastly different internal structures depending on the value of the skyrmion number Q_s.

A nice example of this statement is the so-called coreless vortex in a ferromagnetic condensate with $n_F = 0$.[3] Imposing $n_F = 0$ in (8.41), the circulation remains equal to the integrated skyrmion density

$$\frac{1}{2\pi}\oint \mathbf{v}_F \cdot d\mathbf{r} = \frac{1}{2\pi}\int \mathbf{m}\cdot\left(\frac{\partial \mathbf{m}}{\partial x}\times\frac{\partial \mathbf{m}}{\partial y}\right)dxdy, \tag{8.43}$$

regardless of the choice of the integration loop. As the loop size shrinks, so does the value of the integral on the right, and therefore the value on the left-hand side also shrinks. Unlike the core of the U(1) vortex, however, the core region here is not accompanied by a diverging velocity. This kind of structure is commonly known as the coreless vortex, as there is no need to deplete the density at the core to prevent the kinetic energy catastrophe.

As to writing down an explicit structure of the coreless vortex, one needs to keep in mind that the velocity field is

$$\mathbf{v}_F = \nabla\theta - \cos\beta\nabla\alpha, \tag{8.44}$$

and that the integral $\oint \mathbf{v}_F \cdot d\mathbf{r}$ must vary continuously from a value of 4π for an infinitely large loop, down to zero as the loop size shrinks. One can meet both these boundary conditions by choosing $\nabla\theta = \nabla\alpha$, and then choosing α to coincide with the cylindrical angle φ, $\nabla\alpha = \nabla\varphi = \hat{\varphi}/r$. The velocity field then becomes

$$\mathbf{v}_F = \frac{1-\cos\beta}{r}\hat{\varphi}. \tag{8.45}$$

At large distance $\cos\beta$ must approach -1, to have two units of circulation as dictated by (8.43), and approach $+1$ at the origin to prevent the divergence in the velocity.

[3]The idea of a coreless vortex was originally proposed in the context of the A-phase of ^{3}He and is known as the Anderson-Toulouse-Chechetkin vortex. Mathematically, this is just a two-dimensional skyrmion spin texture. Another kind of topological defect known as the Mermin-Ho vortex has half the winding number of the Anderson-Toulouse-Chechetkin vortex. Such spin structure is also known as the meron.

Attentive readers must no doubt recall that a similar boundary condition has been encountered before. In fact, $\mathbf{v}_F$ equals twice the emergent gauge potential $\mathbf{a}_s$ of a single skyrmion structure [cf. (3.83)].

To summarize, a good variational wavefunction for the coreless ferromagnetic vortex is given by

$$\psi_F = e^{i\varphi}\eta[\mathbf{d} = \mathbf{d}_s], \tag{8.46}$$

where the $\mathbf{d}_s$ vector traces the two-dimensional skyrmion texture. The topologically trivial, coreless vortex may cost less energy than its singular counterpart to generate, since the kinetic energy catastrophe can be avoided.

Obviously, one can go on indefinitely with this sort of discussions for higher-component spinors, facing increased challenges at each level. At the same time, the connection to the skyrmions we know will become increasingly more remote. As such, this is a good place to stop and refer interested readers to dedicated reviews such as Ref. [5].

References

1. Herbut, I.F., Oshikawa, M.: Stable skyrmions in spinor condensates. Phys. Rev. Lett. **97**, 080403 (2006)
2. Lee, D.H., Kane, C.L.: Boson-vortex-skyrmion duality, spin-singlet fractional quantum Hall effect, and spin-$\frac{1}{2}$ anyon superconductivity. Phys. Rev. Lett. **64**, 1313 (1990)
3. Ho, T.L.: Spinor Bose condensates in optical traps. Phys. Rev. Lett. **81**, 742 (1998)
4. Salomaa, M.M., Volovik, G.E.: Half-quantum vortices in superfluid ^{3}He-A. Phys. Rev. Lett. **55**, 1184 (1985)
5. Kawaguchi, Y., Ueda, M.: Spinor Bose-Einstein condensate. Phys. Rep. **520**, 253 (2012)

Index

J.H. Han, *Skyrmions in Condensed Matter*, Springer Tracts
in Modern Physics 278, https://doi.org/10.1007/978-3-319-69246-3

GPSR Compliance
The European Union's (EU) General Product Safety Regulation (GPSR) is a set of rules that requires consumer products to be safe and our obligations to ensure this.

If you have any concerns about our products, you can contact us on

ProductSafety@springernature.com

In case Publisher is established outside the EU, the EU authorized representative is:

Springer Nature Customer Service Center GmbH
Europaplatz 3
69115 Heidelberg, Germany

www.ingramcontent.com/pod-product-compliance
Ingram Content Group UK Ltd.
Pitfield, Milton Keynes, MK11 3LW, UK
UKHW021834270726
14058UKWH00001B/143

* 9 7 8 3 3 1 9 8 8 7 4 1 8 *